Technikzukünfte, Wissenschaft und Gesellschaft / Futures of Technology, Science and Society

Series Editor
Armin Grunwald, ITAS, Karlsruhe Institute of Technology, Karlsruhe, Germany
Reinhard Heil, ITAS, Karlsruhe Institute of Technology, Karlsruhe, Germany
Christopher Coenen, ITAS, Karlsruhe Institute of Technology, Karlsruhe, Germany

Diese interdisziplinäre Buchreihe ist Technikzukünften in ihren wissenschaftlichen und gesellschaftlichen Kontexten gewidmet. Der Plural „Zukünfte" ist dabei Programm. Denn erstens wird ein breites Spektrum wissenschaftlich-technischer Entwicklungen beleuchtet, und zweitens sind Debatten zu Technowissenschaften wie u. a. den Bio-, Informations-, Nano- und Neurotechnologien oder der Robotik durch eine Vielzahl von Perspektiven und Interessen bestimmt. Diese Zukünfte beeinflussen einerseits den Verlauf des Fortschritts, seine Ergebnisse und Folgen, z. B. durch Ausgestaltung der wissenschaftlichen Agenda. Andererseits sind wissenschaftlich-technische Neuerungen Anlass, neue Zukünfte mit anderen gesellschaftlichen Implikationen auszudenken. Diese Wechselseitigkeit reflektierend, befasst sich die Reihe vorrangig mit der sozialen und kulturellen Prägung von Naturwissenschaft und Technik, der verantwortlichen Gestaltung ihrer Ergebnisse in der Gesellschaft sowie mit den Auswirkungen auf unsere Bilder vom Menschen.

This interdisciplinary series of books is devoted to technology futures in their scientific and societal contexts. The use of the plural "futures" is by no means accidental: firstly, light is to be shed on a broad spectrum of developments in science and technology; secondly, debates on technoscientific fields such as biotechnology, information technology, nanotechnology, neurotechnology and robotics are influenced by a multitude of viewpoints and interests. On the one hand, these futures have an impact on the way advances are made, as well as on their results and consequences, for example by shaping the scientific agenda. On the other hand, scientific and technological innovations offer an opportunity to conceive of new futures with different implications for society. Reflecting this reciprocity, the series concentrates primarily on the way in which science and technology are influenced social and culturally, on how their results can be shaped in a responsible manner in society, and on the way they affect our images of humankind.

More information about this series at http://www.springer.com/series/13596

Andreas Lösch · Armin Grunwald ·
Martin Meister · Ingo Schulz-Schaeffer
Editors

Socio-Technical Futures Shaping the Present

Empirical Examples and Analytical Challenges

 Springer VS

Editors
Andreas Lösch
Institut für Technikfolgenabschätzung
und Systemanalyse (ITAS)
Karlsruher Institut für Technologie
Karlsruhe, Germany

Armin Grunwald
Institut für Technikfolgenabschätzung
und Systemanalyse (ITAS)
Karlsruher Institut für Technologie
Karlsruhe, Germany

Martin Meister
Institut für Soziologie
Technische Universität Berlin
Berlin, Germany

Ingo Schulz-Schaeffer
Institut für Soziologie
Technische Universität Berlin
Berlin, Germany

ISSN 2524-3764   ISSN 2524-3772   (electronic)
Technikzukünfte, Wissenschaft und Gesellschaft / Futures of Technology, Science and
Society
ISBN 978-3-658-27154-1   ISBN 978-3-658-27155-8   (eBook)
https://doi.org/10.1007/978-3-658-27155-8

Springer VS

This Springer VS imprint is published by the registered company Springer Fachmedien
Wiesbaden GmbH part of Springer Nature.
The registered company address is: Abraham-Lincoln-Str. 46, 65189 Wiesbaden, Germany

Contents

Editors and Contributors

About the Editors

Andreas Lösch Institute for Technology Assessment and Systems Analysis, Karlsruhe Institute of Technology, Karlsruhe, Germany

Armin Grunwald Institute for Technology Assessment and Systems Analysis, Karlsruhe Institute of Technology, Karlsruhe, Germany

Martin Meister Department of Sociology, TU Berlin, Berlin, Germany

Ingo Schulz-Schaeffer Department of Sociology, TU Berlin, Berlin, Germany

Contributors

Stefan C. Aykut Department of Socioeconomics, University of Hamburg, Hamburg, Germany

Michael J. Bernstein School for the Future of Innovation in Society, Arizona State University, Tempe, USA

Knud Böhle Institute for Technology Assessment and Systems Analysis, Karlsruhe Institute of Technology, Karlsruhe, Germany

Inge Böhm Mannheim, Germany

Christopher Coenen Institute for Technology Assessment and Systems Analysis, Karlsruhe Institute of Technology, Karlsruhe, Germany

Sascha Dickel Institute for Sociology, Johannes Gutenberg University, Mainz, Germany

Paulina Dobroc Institute for Technology Assessment and Systems Analysis, Karlsruhe Institute of Technology, Karlsruhe, Germany

Tess Doezema School for the Future of Innovation in Society, Arizona State University, Tempe, USA

Arianna Ferrari Department Strategy and Content, Futurium gGmbH, Berlin, Germany

Daniela Fuchs Institute of Technology Assessment, Austrian Academy of Sciences, Vienna, Austria

Bruno Gransche Institute of Advanced Studies - FoKoS, University of Siegen, Siegen, Germany

Armin Grunwald Institute for Technology Assessment and Systems Analysis, Karlsruhe Institute of Technology, Karlsruhe, Germany

Alexandra Hausstein Institute of Technology Futures, Karlsruhe Institute of Technology, Karlsruhe, Germany

Reinhard Heil Institute for Technology Assessment and Systems Analysis, Karlsruhe Institute of Technology, Karlsruhe, Germany

Dirk Hommrich Headoffice, German Council for Scientific Information Infrastructures, Göttingen, Germany

Karen Kastenhofer Institute of Technology Assessment, Austrian Academy of Sciences, Vienna, Austria

Lauren Withycombe Keeler School for the Future of Innovation in Society, Arizona State University, Tempe, USA

Jörg Knieling Institute of Urban Planning and Regional Development, HafenCity Universität Hamburg, Hamburg, Germany

Kornelia Konrad Department of Science, Technology and Policy Studies, University of Twente, Enschede, The Netherlands

Andreas Lösch Institute for Technology Assessment and Systems Analysis, Karlsruhe Institute of Technology, Karlsruhe, Germany

Martin Meister Department of Sociology, TU Berlin, Berlin, Germany

Uli Meyer Munich Center for Technology in Society, Technical University of Munich, Munich, Germany

Andreas Mitzschke Faculty of Arts and Social Sciences, Maastricht University, Maastricht, The Netherlands

Gustaf Nelhans University of Borås, Swedish School of Library and Information Science, Borås, Sweden

Jan Nolin University of Borås, Swedish School of Library and Information Science, Borås, Sweden

Alfred Nordmann Institute of Philosophy, Technical University of Darmstadt, Darmstadt, Germany

Nasrine Olson University of Borås, Swedish School of Library and Information Science, Borås, Sweden

Martin Sand Department of Values, Technology and Innovation, TU Delft, BX Delft, The Netherlands

Petra Schaper-Rinkel Center Innovation Systems & Policy, Austrian Institute of Technology, Vienna, Austria

Dirk Scheer Institute for Technology Assessment and Systems Analysis, Karlsruhe Institute of Technology, Karlsruhe, Germany

Christoph Schneider Institute for Technology Assessment and Systems Analysis, Karlsruhe Institute of Technology, Karlsruhe, Germany

Ingo Schulz-Schaeffer Department of Sociology, TU Berlin, Berlin, Germany

Cynthia Selin School for the Future of Innovation in Society & School of Sustainability, Arizona State University, Tempe, USA

Philipp Späth Institute of Environmental Social Sciences and Geography, University of Freiburg, Freiburg, Germany

Helge Torgersen Institute of Technology Assessment, Austrian Academy of Sciences, Vienna, Austria

Alexander Wentland Munich Center for Technology in Society, Technical University of Munich, Munich, Germany

Silvia Woll Institute for Technology Assessment and Systems Analysis, Karlsruhe Institute of Technology, Karlsruhe, Germany

Go Yoshizawa Division of AFI, Oslo Metropolitan University, Oslo, Norway

Introduction: Socio-Technical Futures Shaping the Present

Andreas Lösch, Armin Grunwald, Martin Meister and Ingo Schulz-Schaeffer

1   Shaping the Present Through Futures

Socio-technical futures like visions or scenarios, that is the more or less systematic imagination of the interplay of future technologies with future society, have played an important role in Science and Technology Studies (STS) and Technology Assessment (TA) for decades. For a long time, the debates on and the analysis of these socio-technical futures have focused on the plausibility and legitimacy of these futures as anticipations, both on the technical and on the societal side. Only recently, the focus has shifted to investigating how socio-technical images of the future influence the present, that is, how they contribute to present processes of policymaking, societal debate and technology development, equally. This volume presents papers that focus on the question how socio-technical images of the future shape present processes of innovation and transformation.

A. Lösch (✉) · A. Grunwald
Institute for Technology Assessment and Systems Analysis, Karlsruhe Institute of Technology, Karlsruhe, Germany
e-mail: andreas.loesch@kit.edu

A. Grunwald
e-mail: armin.grunwald@kit.edu

M. Meister · I. Schulz-Schaeffer
Department of Sociology, TU Berlin, Berlin, Germany
e-mail: martin.meister@tu-berlin.de

I. Schulz-Schaeffer
e-mail: schulz-schaeffer@tu-berlin.de

© Springer Fachmedien Wiesbaden GmbH, part of Springer Nature 2019
A. Lösch et al. (eds.), *Socio-Technical Futures Shaping the Present*,
Technikzukünfte, Wissenschaft und Gesellschaft / Futures of Technology,
Science and Society, https://doi.org/10.1007/978-3-658-27155-8_1

The influence of images of the future on technology development and technology assessment in the present is not a completely new topic to Science and Technology Studies (STS) and to Technology Assessment (TA). A fundamental presupposition of STS is that any imagination of new technologies is necessarily coupled with the shaping or even inventing of novel kinds of users and usages, of images of altered social relations or even new societal regimes. However, within the framework of social construction of technology and society, those images of socio-technical futures were only conceptualized as one among many devices used by social groups in present negotiation processes, while the emphasis was on reconstructing these negotiations. Almost from its beginning, TA developed methodologies to specify images of plausible, but diverging versions of future socio-technical relations in the classical scenario approach. Sharply distinguished from forecasts, these scenarios were used to create a knowledge base for present political and societal decisions between alternative paths of development. As devices for supporting present decision processes, emphasis was on the plausibility of the constructed scenarios to reveal the consequences of contemporary decisions for avoiding the worst-case scenario and for making the achievement of the best-case scenario more likely.

In both strands of research, the topic of how socio-technical images of the future shape present processes of innovation and transformation has been there for quite a long time. However, there have been only few attempts to address the topic directly, conceptually or empirically. Neither in STS nor in TA much attention has been drawn to scrutinize in detail the influence of socio-technical futures in the present, e.g. their influence on orientations and decisions in concrete technology developments and broader innovation processes, or their influence on contemporary societal debates. This also holds true for historical studies: Not much evidence has been gathered about how past socio-technical images of the future influenced past negotiations and processes of decision-making. Furthermore, there are only a few attempts to elaborate on the constitutive interplay between imagined and factual socio-technical developments.

Addressing the question of how socio-technical images of the future shape present processes of innovation and transformation is to ask questions like the following:

- What kinds of future concepts are at play? What role do they play for the shaping of present practices and processes?
- How could we differentiate these future concepts according to their functions in social processes (e.g., policy and public debates, technology development and innovation or technology governance)?

- What patterns of guidance through these concepts can be generalized from empirical case studies?
- What kinds of interfaces between present and future so these future concepts provide?
- How to use the insights in different effects of socio-technical futures for technology assessments, foresight, technology development, and in technology governance?
- Moreover, as an overarching question: What are ways to conceptualize the shaping of the present by socio-technical futures?

The contributions to this volume attempt to provide answers to these questions. Their point of departure is quite different: Some start from empirical case studies and generalize specific findings in the light of these questions. Other contributions tackle conceptual questions from the outset. Despite these differences, all of these contributions share a common understanding of the basic problem at hand. It is clear that we cannot know the future precisely, but on the other hand imagining this future is influential for the present state of affairs. Therefore, the basic assumption of this volume is that there is a negotiation between the present and possible futures, a debate that is empirically detectable and in conceptual respect constitutive in both the realm of public discourse and policymaking, but also in concrete research and development and engineering endeavors.

2 Some Precursors

The idea that socio-technical images of the future shape present processes of technology development, innovation and transformation has influential precursors both in STS and in TA which we briefly want to remind of.

In the phase of formation of STS in the 1980s and 1990s, some research pointed to the performative role of imaginaries and future imagination as one element in the social construction of technology (Bijker et al. 1987) and, in the same instance, the shaping of society at large (Bijker and Law 1992). An influential example for this is the concept of material technology embodying engineers "scripts" that prescribe future use of technology (Akrich 1992). According to the author herself, these scripts have the form of concrete socio-technical scenarios that describe future use and users in some detail. Another example is the German "Leitbild" approach (Dierkes et al. 1996) that gathered some evidence for the influence of larger socio-technical visions for technology-related decision-making on the larger societal scale. In the same line, the concept of "guiding visions"

(Grin and Grunwald 2000) focused on the integrative and coordinative function of such future imaginations on the different stages of processes of technology development and innovation.

On a more general level, the "sociology of expectations" (Brown and Michael 2003) analyzed how "contested futures" (Brown et al. 2000) or stabilizations and destabilizations of "collective expectations" (Konrad 2006) influence the course of innovations or the formation of a new technology (e.g. Selin 2007, on nanotechnology). Oriented by conceptual models as the "promise requirement cycle" (van Lente 1993) or the "translation model" (Callon 1995), these scholars highlighted that expectation dynamics are a crucial factor to explain why processes of technological innovations often differ from the typical stages and curves depicted in classical innovation research and management (Borup et al. 2006). On a philosophical, that is even more general level conceptualizations of the role of different temporalities of how "future matters" (Adam and Groves 2007) provide a theoretical background for analyzing the shaping of the present (or past presents) by futures.

In the multi-level approach of innovation (Geels and Schot 2010) and the related concept of "transition research" (e.g. Geels 2005; Nill and Kemp 2009) visions were identified as crucial elements besides others, which allow to transfer inventions out of their niches to the level of established technological regimes, thus transforming the latter. Within this strand of conceptualizing innovation, some research was done to conceptualize scenarios of the future as concrete "articulations of visions and expectations" (Geels 2005, p. 366; see also Rotmans and Loorbach 2010) or as "promising technology" (van Lente 1993) that strongly influences present developments by strategically imagining promising future states of affairs, both on the technological and the societal side.

So while the conceptual foundations have been there for quite a long time, the potential of addressing imagined socio-technical futures as an important factor for influencing concrete present technology-related developments and decisions has only in rare cases been spelled out conceptually as well as empirically. The well-known concept of "socio-technical imaginaries" (Jasanoff and Kim 2015) focusses on the role of these imaginaries in national or transnational controversies about technologies and in particular on their political consequences (see also Marcus 1995). The likewise well-known research drawing on Woolgar (1991) concentrated on the question how much the possible activities of users are constrained by "configuring" these users as imagined entities in the development process.

In STS, there have been only few studies that directly focus on the role imagined socio-technical futures play as an important factor for influencing concrete present technology-related social change and technology developments directly.

Examples are the study of Fujumura (2003) on genome scientists, who use images of the future as part of their "sociocultural entrepreneurship" to enforce their version of biology and especially applied biotechnology. Another example is the study of Hyysalo (2006) on the role of diverging representations of future use of health care technologies by different professionals in this sector, which makes these diverging professional imaginations of future states of affairs the battleground for the enforcement of a distinguished version of technical supported elderly care. Only recently, the empirical investigation of the role of future imaginations in present processes gained more attention. The goal of this volume is to foster debate and conceptual mutual reference of these approaches and empirical findings on the present role of imagined socio-technical futures for the state of the art in STS and the sociology of technology, respectively.

STS approaches, which reflect the shaping impacts of future imaginations on practices and processes of sociotechnical innovation and transformation, were mainly orientated retrospectively. That is, they provided evidence for the shaping functions and effects of sociotechnical futures based on retrospective case studies. In contrast to this focus, TA research is orientated prospectively. Corresponding to the aim and mission of TA to provide science based knowledge for parliamentary or policy advice or consultations of other societal actors concerning the future impacts of not yet established but upcoming or transforming TA research cannot mainly focus on past processes. TA has to generate knowledge for action in confrontation with just emerging or ongoing processes, without knowing the outcome of the processes (Grunwald 2018). Correspondingly, the perspective of TA in its major areas can be characterized as a consequentialist approach. Following this consequentialist mode of operation, TA aims to create knowledge about consequences of the technology under consideration in its presumed social context, has then to assess this knowledge with respect to normative criteria or by applying participatory approaches, and can finally use the assessment results as source for providing orientation to today's deliberations and decision-making processes. Usually, TA following its consequentialist scheme is regarded as a reflective approach to shape the future according to social values or political goals by prospective reasoning (e.g. Guston and Sarewitz 2002). The idea of shaping the future by present reasoning also dominates the related field of Responsible Research and Innovation (RRI, cp. Owen et al. 2013; van den Hoven et al. 2014). It found its clearest but contested expression in the proposal that RRI should aim at "achieving the right impacts" (von Schomberg 2013) which puts RRI in close neighborhood to advanced types of planning.

The issue of our book, the shaping impacts of sociotechnical imaginations of futures in present technological developments, practices of innovation and transformations and political and public debates was and still is not a central perspective in TA. However, the arrival of "new and emerging technologies" (NEST) such as nanotechnology, human enhancement, synthetic biology, robotics, and quantum technologies posed a particular challenge for the mainstream consequentialist perspective of TA. The demand for grasping the shaping impacts of future imaginations—in line with the social constructivism in STS—was growing in confrontation with the phenomena of the NEST starting with the field of nanotechnology (e.g. Kaiser et al. 2010; Grunwald 2013). Among the major challenges was the observation that techno-visionary futures were proposed and intensively discussed without any chance of relating them with sound epistemological knowledge. Utopian and dystopian futures, paradise-like and apocalyptic expectations, bright and dark images of the future dominated public and scientific debate, e.g. in the early years of the nanotechnology debate (e.g. Coenen 2010) and motivated policymakers to establish research funding programs (e.g. NSTC 1999) as well as precautionary measures (e.g. in European legislation). The irritation to TA was that those visionary futures had visible impact on public debate and policy-making although it was not possible to provide policy advice on these issues within its familiar consequentialist scheme. The reason simply was lack of knowledge about future developments to be used as epistemic basis for TA assessment processes.

This experience motivated new types of reasoning. Vision Assessment (Grunwald 2009; Lösch et al. 2016, see also the white paper in this volume) was developed exactly to better understand visionary communication and to attempt to extract orientation for society and policy-makers out of this knowledge by applying concepts beyond consequentialism (Grunwald 2014; Nordmann 2014). It gave rise to adding a further mode of orientation to the traditional ones of prediction and scenario-building mentioned above: the hermeneutic mode renouncing on any attempt to anticipate future developments but rather taking narratives of the future as expressions in current and ongoing communication (Grunwald 2016). In both approaches, Vision Assessment as well as hermeneutic TA, one of the main questions is how narratives of the future shape the present.

Vision Assessment and hermeneutic TA are transformations at the "meeting points" of TA and STS (see, e.g. studies on the role of sociotechnical futures in the case of the energy transition from Lösch and Schneider 2016; Engels and Münch 2015). Especially evident are these shifts in the ongoing extensions in the modes and concepts of anticipatory governance (e.g. Barben et al. 2008) towards the reflection of the use of the present impacts of sociotechnical futures (e.g. Konrad et al. 2017). In TA, there is a growing reflection on the meanings and shaping roles in present affairs (e.g., practices of research policy, public perception, real-life experiments, governance) observable. But the consequentialist scheme is still dominant and important for all of the future related reasoning in which TA is involved.

This volume puts the growing awareness and research perspectives in STS and TA in the center by assembling various analytical gazes and cases-studies, which try to analytically grasp how envisioning and modeling possible socio-technical futures influences, steers, frames or even enables practices and processes of socio-technical innovation and transformation in the present. The focus are the questions on the ways, how different types of socio-technical futures are shaping the present.

3 Arrangement of Chapters

3.1 Analytical Concepts, Different Kinds and Patterns of Socio-Technical Futures

Armin Grunwald asks what it does mean to develop and employ socio-technical futures in order to shape the future. He argues that since at any time we are only able to intervene into the present, we cannot shape the future directly. We can only intervene into present constellations and expect or hope that these interventions will lead to future developments in good resonance with the goals and purposes related to the respective intervention. Thus, our present imaginations of socio-technical futures, though they do not represent future presents but only present futures, nevertheless may have an influence on the future by orienting our present future-oriented debates, decisions and actions. For this reason, imagined socio-technical futures remain to be an important subject for technology assessment. However, viewing socio-technical futures as present futures calls for a hermeneutic extension of technology assessment that analyzes what the pictures of the future tell us about us today rather than about the future.

Ingo Schulz-Schaeffer and *Martin Meister* regard situational scenarios and prototype scenarios as socio-technical futures that shape present socio-technical developments in a specific way. The process of scenario-building is understood as integrating heterogeneous components, some of which are already existing and some of which still have to be created in the future. The necessity to integrate these present and future components allows considering scenario-building as a kind of negotiation between the present and imagined futures. The authors focus on how the envisaged contexts of use are represented in these negotiations. They base their analysis on empirical research in the field of ubiquitous engineering.

The analysis provided allows for a distinction between different forms of representation and for an assessment of their respective negotiation power.

Drawing on the concept of "vision assessment", *Andreas Lösch, Reinhard Heil,* and *Christoph Schneider* address "visionary practices shaping power constellations" in three fields of technology development that at the moment are highly promising, but also highly contested: Big Data, Smart Grid and FabLabs. Oriented to their analytical gaze on "visions as socio-epistemic practices", they conceptualize visions as to entail discursive processes and technology related practices, as well as material and organizational aspects. They demonstrate the differences between these cases with respect to their central claim about different ascriptions of responsibility in this process. This calls for a more detailed analysis of present power relations in socio-technical innovation processes.

3.2 Socio-Technical Futures in Different Processes of Change

Tess Doezema examines how U.S.-based visions of biotechnological potential, undergirded by narratives of biological and economic progress, contribute to shaping technological innovation, policy, and politics at the international level. She argues that the invocation of "science-based" markets and policies is put to work in diverse instances to create conditions of potentiality for a particular vision of future good, the bioeconomy vision. This contribution examines the present roles and impacts of future visions in the context of techniques for the creation of international markets for biotechnologies, as well as the modes of measurement and obfuscation leveraged in their validation and discursive construction as universally beneficial, placeless, and disembodied technologies. This calls for a stronger awareness on the interdependences of global and local dimensions in studying the shaping impacts of visions.

Uli Meyer shows how socio-technical futures can contribute to the formation of issue-based fields with the envisioned future at its core. He argues that in such cases organizations orient and coordinate their activities in pursuit of the envisioned future. In his contribution, he analyzes the example of the emerging field of Industry 4.0 in Germany to understand why, how and under what circumstances such imagined futures tend to emerge, diffuse, and stabilize. Conceptually, Meyer complements considerations from the "sociology of expectations" with Karl Weick's concept of sensemaking. He argues that socio-technical futures are used as a sensemaking mechanism that facilitates the emergence of interorganizational networks and thus leads to the formation of issue-based fields.

Combining the theoretical frames of the "social construction of technology" (Bijker et al. 1987) and "sociotechnical imaginaries" (Jasanoff and Kim 2015) *Andreas Mitschke* explores the enduring controversy about genetically modified crops in India. He analyses how different interpretations of past agricultural development play important roles in the construction of contested sociotechnical futures with or without transgenic crops. By focusing on the temporal dimension of sociotechnical imaginaries, he sheds light on how the different positions in the controversy are shaped by different interpretations of India's agricultural past. His case provides evidence on how past presents and presents are interrelated and constitutive for an analysis on the role of future imaginations in the shaping of technological controversies. This calls for a more intensive reflection of the constitutive temporalities of sociotechnical futures.

In their contribution, *Philipp Späth* and *Jörg Knieling* make use of the concept of Anderson (2010) who distinguishes a style of premediation, practices of performing and a logic of preparedness as fundamentally different styles of TA- or mediation processes. Investigating drastical changes in framing the negotiation on Hamburg as a "Smart City", they show different ways of "dealing with future in Hamburg". Their basic claim is that a logic of planning, characterized by the styles of foresight and prediction, by practices of calculating and the logic of precaution, was replaced or at least complemented and challenged by an attitude of experimenting towards real-time management.

3.3 Intervening into the Present Through Prospective Reasoning

Go Yoshizawa starts his analysis by recapitulating the consequentialist approach dominating TA's practice. TA has been employed by policymakers and practitioners to formulate and deliberate policy and social agendas by anticipating plausible futures for some fifty years. The author proposes a reflexive hermeneutics approach and illustrates it at the occasion of synthetic biology. The case study allows the conclusion that future-oriented conceptions of the 'precautionary' and 'proactionary' intertwine with general biological and bio-technological concepts such as biosafety, biosecurity, bioeconomy, and biodiversity. The author concludes that these concepts already have been shaping the present by affecting perception of possible futures.

The point of departure of the chapter authored by *Inge Böhm, Silvia Woll* and *Arianna Ferrari* is the diagnosis that the current production of meat and the high meat consumption have negative effects on humans, animals and the environment. They regard a sustainable reorientation of mass production and mass consumption as not possible. Innovators declare in vitro *meat* as a technological solution of the problem. Replacing meat from animals by in vitro meat is seen as an innovation, which could solve the problem without giving up having meat in our food. The authors present results from expert and stakeholder interviews after having confronted them with the vision of the in vitro meat innovation. Finally, insight in current visions and imaginations of possible futures are provided demonstrating particular challenges also beyond in vitro meat itself.

The contribution of *Jan Nolin, Gustaf Nelhans, Nasrine Olson* problematizes that the involvement of a multitude of stakeholders in Constructive Technology Assessment approaches for deliberating on involving technology often lack deeper understanding of the concepts which mediate the discussion and are thus confused by concepts regarding "technology rich futures". For the field of emerging ICTs they provide insights on ten influential concepts and their dynamics that have played an important role in the ICT sector during the 2000's based on bibliometric data. Their approach and analysis supports the understanding in TA projects on how concepts used to describe futures of ICTs influence the present technological development and societal changes. This calls for a stronger awareness on the impacts of concepts and concept-variations in research of their roles in innovations and transformations.

Lauren Withycombe Keeler, Michael J Bernstein, and *Cynthia Selin* explore participatory scenario construction as a means to productively disrupt status-quo imaginaries. The Future of Wastewater Sensing, a participatory scenario study, is presented as a case example to inform sustainability-oriented responsible research and innovation. Discourses around innovation often unreflexively assume positive progress and the inevitable contribution of new technologies to the betterment of society. Little attention is paid to issues of sustainability and the complex ways that societal arrangements and sociotechnical regimes are intermingled. Participatory scenario construction is a way to utilize tools of foresight to raise the capacity of actors in innovation processes to consider alternative framings of progress and challenge the status quo.

3.4   White Paper on Technology Assessment and Socio-Technical Futures

The fourth part of the volume consist of the programmatic and collective white paper "Technology Assessment of socio-technical futures" written by 23 authors.

The paper outlines how present significance and effects of socio-technical futures in present technological research and innovation processes can be conceived of and analyzed. It discusses, why TA should analyze sociotechnical futures, how such analyzes can grasp the societal conditions that are expressed in the imagined futures and how these become effective in processes of technology development, communication, decision making etc. Finally, it elaborates which self-reflexive positioning or possible realignment of TA are needed, caused by the increased concern with assessing and even co-producing sociotechnical futures by TA. The aim of the paper is to sensitize colleagues in the TA community in a broad sense for the topic and its challenges, to consolidate discussions and to provide theoretical and methodical suggestions for research in TA and related advisory practices with respect to sociotechnical futures. The origin of this paper has been the workshop "The present of technological futures-theoretical and methodical challenges for Technology Assessment" (March 2016, Karlsruhe), in which all of the paper's authors participated. The contents of this white paper are preliminary results that shall initiate and guide further discussions.

4 Background of the Volume

The idea for this volume was arising during the track "Socio-technical Futures Shaping the Present-Analytical Challenges in STS and Technology Assessment", which we editors organized at the 4S/EASST-Conference, in Barcelona in August/September 2016. The volume assembles papers from the track, additional invited papers from selected authors, and the English version of the white paper in order to complete the spectrum of research topics to round up this volume. Finally, the editors thank Philipp Frey from ITAS for his critical input on some chapters and his assistance in the editorial tasks during the preparation of the manuscripts for the publisher.

References

Adam, B., & Groves, C. (2007). *Future matters: Action, knowledge, ethics.* Leiden: Brill.
Akrich, M. (1992). The de-scription of technical objects. In W. Bijker & J. Law (Eds.), *Shaping technology, building society* (pp. 205–224). Cambridge: MIT Press.
Anderson, B. (2010). Preemption, precaution, preparedness: Anticipatory action and future geographies. *Progress in Human Geography, 34*(6), 777–798.

Barben, D., Fisher, E., Selin, C., & Guston, D. (2008). Anticipatory governance of nano-technology: Foresight, engagement, and integration. In E. Hachett, O. Amsterdamska, M. Lynch, & J. Wajcman (Eds.), *The handbook of science and technology studies* (3rd ed., pp. 979–1000). Cambridge: MIT Press.

Bijker, W. E., & Law, J. (Eds.). (1992). *Shaping technology/building society.* Cambridge: MIT Press.

Bijker, W. E., Hughes, T. P., & Pinch, T. J. (Eds.). (1987). *The social construction of technological systems. New directions in the sociology and history of technological systems.* Cambridge: MIT Press.

Borup, M., Brown, N., Konrad, K., & van Lente, H. (2006). The sociology of expectations in science and technology. *Technology Analysis & Strategic Management, 18*(3–4), 285–298.

Brown, N., & Michael, M. (2003). A sociology of expectations: Retrospecting prospects and prospecting retrospects. *Technology Analysis & Strategic Management, 15*(1), 4–18.

Brown, N., Rappert, B., & Webster, A. (Eds.). (2000). *Contested futures: A sociology of prospective techno-science.* Burlington: Ashgate.

Callon, M. (1995). Four models for the dynamics of science. In S. Jasanoff, G. E. Markle, J. C. Petersen, & T. Pinch (Eds.), *Handbook of science and technology studies* (pp. 29–63). Thousand Oaks: Sage.

Coenen, C. (2010). Deliberating visions: The case of human enhancement in the discourse on nanotechnology and convergence. In M. Kaiser, M. Kurath, S. Maasen, & C. Rehmann-Sutter (Eds.), *Governing future technologies. Nanotechnology and the rise of an assessment regime* (pp. 73–87). Dordrecht: Springer.

Dierkes, M., Hoffman, U., & Marz, L. (1996). *Visions of technology. Social and institutional factors shaping the development of new technologies.* Frankfurt a. M.: Campus.

Engels, F., & Münch, A. V. (2015). The micro smart grid as a materialised imaginary within the German energy transition. *Energy Research & Social Science, 9,* 35–42. https://doi.org/10.1016/j.erss.2015.08.024.

Fujimura, J. H. (2003). Future imaginaries, genome scientists as sociocultural entrepreneurs'. In A. H. Goodman, D. Heath, & S. M. Lindee (Eds.), *Genetic nature/culture* (pp. 176–199). Los Angeles: University of California Press.

Geels, F. W. (2005). *Technological transitions and system innovations. A co-evolutionary and socio-technical analysis.* Cheltenham: Eigar.

Geels, F. W., & Schot, J. (2010). The dynamics of transitions: A socio-technical perspective. In J. Grin, J. Rotmans, J. Schot, & F. W. Geels (Eds.), *Transitions to sustainable development: New directions in the study of long term transformative change* (pp. 11–101). New York: Routledge.

Grin, J., & Grunwald, A. (Eds.). (2000). *Vision assessment: Shaping technology in 21st century society. Towards a repertoire for technology assessment.* Berlin: Springer.

Grunwald, A. (2009). Vision assessment supporting the governance of knowledge-The case of futuristic nanotechnology. In G. Bechmann, V. Gorokhov, & N. Stehr (Eds.), *The social integration of science. Institutional and epistemological aspects of the transformation of knowledge in modern society* (pp. 147–170). Berlin: Edition Sigma.

Grunwald, A. (2013). Modes of orientation provided by futures studies: Making sense of diversity and divergence. *European Journal of Futures Research, 2,* 30. https://doi.org/10.1007/s40309-013-0030-5.

Grunwald, A. (2014). The hermeneutic side of responsible research and innovation. *Journal of Responsible Innovation, 1*(3), 274–291.

Grunwald, A. (2016). *The hermeneutic side of responsible research and innovation.* London: Wiley.

Grunwald, A. (2018). *Technology assessment. From practice to theory.* London: Routledge.

Guston, D., & Sarewitz, D. (2002). Real-time technology assessment. *Technology in Society, 24,* 93–109.

Hyysalo, S. (2006). Representations of use and practice-bound imaginaries in automating the safety of the elderly. *Social Studies of Science, 36*(4), 599–626.

Jasanoff, S., & Kim, S.-H. (2015). *Dreamscapes of modernity. sociotechnical imaginaries and the fabrication of power.* Chicago: Chicago University Press.

Kaiser, M., Kurath, M., Maasen, S., & Rehmann-Sutter, C. (Eds.). (2010). *Governing future technologies; Nanotechnology and the rise of an assessment regime.* Dordrecht: Springer.

Konrad, K. (2006). The social dynamics of expectations: The interaction of collective and actor-specific expectations on electronic commerce and interactive television. *Technology Analysis & Strategic Management, 18*(3–4), 429–444.

Konrad, K., van Lente, H., Groves, C., & Selin, C. (2017). Performing and governing the future in science and technology. In U. Felt, R. Fouché, C. A. Miller, & L. Smith-Doerr (Eds.), *The handbook of science and technology studies* (4th ed., pp. 465–493). Cambridge: MIT Press.

Lösch, A., Böhle, K., Coenen, C., Dobroc, P., Ferrari, A., Heil, R., Hommrich, D., Sand, M., Schneider, C., Aykut, S., Dickel, S., Fuchs, D., Gransche, B., Grunwald, A., Hausstein, A., Kastenhofer, K., Konrad, K., Nordmann, A., Schaper-Rinkel, P., Scheer, D., Schulz-Schaeffer, I., Torgersen, H., & Wentland, A. (2016). Technikfolgenabschätzung von soziotechnischen Zukünften. Discussionpaper 3, Institute for Technology Futures. Karlsruhe: KIT. https://publikationen.bibliothek.kit.edu/1000062676.

Lösch, A., & Schneider, C. (2016). Transforming power/knowledge apparatuses: The smart grid in the German energy transition. *Innovation: The European Journal of Social Science Research, 29*(3), 262–284. https://doi.org/10.1080/13511610.2016.1154783.

Marcus, G. (Ed.). (1995). *Technoscientific imaginaries.* Chicago: University of Chicago Press.

Nill, J., & Kemp, R. (2009). Evolutionary approaches for sustainable innovation policies: From niche to paradigm? *Research Policy, 38,* 668–680.

NSTC, National Science and Technology Council. (1999). *Nanotechnology. Shaping the world atom by atom.* Washington, D.C.: NSTC.

Nordmann, A. (2014). Responsible innovation, the art and craft of future anticipation. *Journal of Responsible Innovation, 1*(1), 87–98.

Owen, R., Bessant, J., & Heintz, M. (Eds.). (2013). *Responsible innovation: Managing the responsible emergence of science and innovation in society.* London: Wiley.

Rotmans, J., & Loorbach, D. (2010). Towards a better understanding of transitions and their governance: A systemic and reflexive approach. In J. Grin, J. Rotmans, J. Schot, & F. W. Geels (Eds.), *Transitions to sustainable development: New directions in the study of long term transformative change* (pp. 105–220). New York: Routledge.

Selin, C. (2007). Expectations and the emergence of nanotechnology. *Science, Technology and Human Values, 32*(2), 196–220.

van den Hoven, J., Doorn, N., Swierstra, T., Koops, B.-J., & Romijn, H. (Eds.). (2014). *Responsible innovation 1: Innovative solutions for global issues*. Dordrecht: Springer.

van Lente, H. (1993). *Promising technology. The dynamics of expectations in technological de-velopment*. Delft: Eburon.

von Schomberg, R. (2013). A vision of responsible research and innovation. In R. Owen, J. Bessant, & M. Heintz (Eds.), *Responsible innovation: Managing the responsible emergence of science and innovation in society* (pp. 51–74). London: Wiley.

Woolgar, S. (1991). Configuring the user: The case of usability trials. In J. Law (Ed.), *A sociology of monsters essays on power, technology and domination* (pp. 58–99). London: Routledge.

Andreas Lösch (Ph.D. and habilitation) is sociologist, senior researcher and head of the research area "knowledge society and knowledge policy" and of the research group on "vision assessment" at the Institute for Technology Assessment and Systems Analysis at the Karlsruhe Institute of Technology.

Armin Grunwald (Ph.D. and habilitation) is Director of the Institute for Technology Assessment and Systems Analysis and Full Professor of Philosophy and Ethics of Technology at the Karlsruhe Institute of Technology. Simultaneously, he heads the Office of Technology Assessment at the German Bundestag.

Martin Meister (Ph.D.) is a research associate at the Department of Sociology, Chair on Sociology of Technology and Innovation at Technical University of Berlin. Besides sociology and technology studies, actual interest is on technology assessment and social robotics.

Ingo Schulz-Schaeffer (Ph.D. and habilitation) is Full Professor and head of the sociology of technology and innovation group at the Institute of Sociology, Technical University of Berlin. He is principal investigator of the DFG cluster of excellence "science of intelligence", and member of the program committee of the DFG priority program "digitalization of working worlds".

Part I
Analytical Concepts, Different Kinds and Patterns of Socio-Technical Futures

Shaping the Present by Creating and Reflecting Futures

Armin Grunwald

Abstract

It is a commonly used rhetoric phrase that we develop ideas how to shape the future and that we shape the future exactly by implementing those ideas. However, what does it mean to "shape the future"? We are only able to intervene into the present, by communication, by action, or by decisions to be made. These interventions then might have consequences for future developments or events. Thus, the phrase should better be reformulated: we do not shape the future itself but we intervene into present constellations and thereby influence future developments more or less indirectly. As far as we use socio-technical futures as orientation to identify appropriate interventions into present constellations-as usually is done by technology assessment—we can speak of futures contributing to shape the present. It means that in present time we create futures supporting us to shape the present.

1   Setting the Stage

It is a commonly used phrase that we develop ideas how to shape the future and that we shape the future by implementing those ideas. While this is the traditional view of planning (Camhis 1979) it also applies to more recent approaches such

A. Grunwald (✉)
Institute for Technology Assessment and Systems Analysis, Karlsruhe Institute of Technology, Karlsruhe, Germany
e-mail: armin.grunwald@kit.edu

© Springer Fachmedien Wiesbaden GmbH, part of Springer Nature 2019   17
A. Lösch et al. (eds.), *Socio-Technical Futures Shaping the Present*,
Technikzukünfte, Wissenschaft und Gesellschaft / Futures of Technology,
Science and Society, https://doi.org/10.1007/978-3-658-27155-8_2

as reflexive governance (Voss et al. 2006). Also the world of NEST (new and emerging sciences and technologies) is full of narratives and pictures why and how NEST should be developed in order to shape the future, e.g. for solving the global energy supply problem, for enhancing human performance, or for designing artificial life.

However, what does it really mean to shape the future? At any time we are only able to intervene into the present, by communication, by action, or by decisions to be made. These interventions then will usually have consequences for future developments or events. Thus, the phrase should better be formulated in the way of the title of this volume: we cannot "shape the future" directly but we only can intervene into present constellations and expect or hope that these interventions will lead to future developments in good resonance with the goals and purposes related to the respective intervention. As far as we develop and use socio-technical futures[1] as orientation to identify appropriate interventions into present constellations we can speak of futures contributing to shape the present. It means that in present time we create futures supporting us to shape the present. This perspective refers to the philosophy of time by Augustine of Hippo (Augustine 397):

> Nec proprie dicitur: tempora sunt tria, praeteritum, praesens et futurum, sed fortasse proprie diceretur: tempora sunt tria, praesens de praeteritis, praesens de praesentibus, praesens de futuris (Augustine 397, p. XI, 20)

> Thus it is not properly said that there are three times, past, present, and future. Perhaps it might be said rightly that there are three times: a time present of things past; a time present of things present; and a time present of things future (Augustine 397, p. XI, 20)

This early observation corresponds to the epistemic immanence of the present (Grunwald 2006): Futures as well as pasts are always parts of the present. There is no possibility to shape the future in the meaning of shaping directly a future present (Luhmann 1998/1992), e.g. of shaping today some elements of the German energy system for the year 2025. Today we cannot intervene into the year 2025 but only into our current present. According to this picture shaping the future only can mean shaping the present with, however, some regard to the (also present!) futures we have in mind.

---

[1]The term 'futures' is used in this chapter to denote *present* imaginations of future developments or events: present futures instead of future presents (according to Luhmann 1998/1992).

In contrast to this Augustinian perspective the widespread talk about "shaping the future" involves a kind of planning optimism. It presupposes that our interventions to the present will lead (at least mostly) to the desired results so that *shaping the future* would only be an abbreviation for *shaping the present and presuming that the consequences of the respective interventions will lead (more or less) to the expected events or developments as soon as the time will have come.*

The Augustinian perspective on time radicalizes further the modest assumptions of the plannability and controllability of the sociotechnical future dominating technology assessment and STS studies since decades. According to this perspective no longer "the future" in the sense of a future present shall be shaped but rather the accent is given to shaping our today's present. By shaping this present processes can be initiated building a processual bridge to future presents—however, not aiming at shaping those futures directly because during the process many alternatives might appear. Thus, this type of reasoning combines anticipatory thinking with the idea of the openness of the future and the existence of many alternative options while proceeding from today's present to some future present.

In the following sections I will unfold the perspective opened up by the Augustinian perspective on time and search for implications for technology assessment and related approaches. First, some observations of real-world impacts of futures on the respective present will be discussed (Sect. 2), in particular taking into consideration examples from NEST (new and emerging sciences and technologies). This section shall illustrate how the present is de facto shaped by futures. Second, the Augustinian perspective will be underpinned by more in-depth considerations of the structure of futures (Sect. 3). Third, consequences in several dimensions will be drawn for technology assessment but also beyond (Sect. 4).

2 Observations: Impacts of NEST Futures on the Present

Making statements about the futures always is an *intervention* in ongoing present-time communication and can have an impact on present-time issues. In particular, in everyday life statements of some developments to occur in the future impact on present-day's issues. For example, the daily weather forecast for tomorrow may influence our plans for hiking activities. Or, take predictions in demographic change, e.g. about the number of pupils and students in ten years are taken as statements about how those numbers will really develop in the coming years.

These statements might be used as information for e.g. planning schoolhouses and educating or hiring teachers and thus making decisions today, independent of whether these prospective statements will prove wrong or false in the time to come. Futures can trigger a turn in a debate and influence decisions, possibly depending on how consistent, plausible, motivating, threatening or fascinating the respective futures are.

While this statement holds generally (e.g. the example of trend extrapolations discussed in Sect. 4) the focus of this chapter is on techno-visionary communication. In the past decades a considerable increase in visionary communication on future technologies and their impacts on society could be observed. In particular, this has been and still is the case in the fields of nanotechnology (Selin 2007; Fiedeler et al. 2010), human enhancement and the converging technologies (Roco and Bainbridge 2002; Grunwald 2007), synthetic biology (Giese et al. 2014), and climate engineering. Visionary scientists, science managers and science authors have put forward far-ranging visions, which have been disseminated by mass media and discussed in science and the humanities (Grunwald 2016) and shall be called techno-visionary futures characterized by (according to Grunwald 2013):

- They refer to a more distant future, some decades ahead, and exhibit revolutionary aspects in terms of technology and in terms of culture, human behaviour, individual and social issues.
- Scientific and technological advances are regarded in a renewed techno-determinist fashion as by far the most important driving force in modern society (technology push perspective).
- Their authors are mostly scientists, science writers and science managers such as Eric Drexler and Ray Kurzweil; but also NGO's and industry are developing and communication visions.
- Milestones and technology roadmaps are provided to demonstrate the feasibility of those futures and to demarcate a difference to narratives Science Fiction.
- Often high degrees of uncertainty are involved with severe controversies as a consequence.

The emergence of this new wave of visionary and partially futuristic communication has provoked renewed interest in the role played by imagined visions of the future. Obviously, there is no distinct borderline between different types of visions communicated in these fields and other imagined futures such as *Leitbilder* or guiding visions which have already been analyzed with respect to their usage in policy advice (Grin and Grunwald 2000). Techno-visionary futures address possible social developments in the light of visionary sciences and their impacts on society at a very early stage of development. As a rule, little if any knowledge is available about how the respective technology is likely to develop, about the products such development may spawn, and about the potential impact of using such products. According to the control dilemma (Collingridge 1980), it is then extremely difficult, if not impossible, to shape technology. Instead, lack of knowledge could lead to a merely speculative debate (e.g. Nordmann 2007), followed by arbitrary communication and conclusions (Grunwald 2016, Chap. 3).

Indeed, one could argue that some of the NEST debates are so speculative that they should better be ignored because of lack of any practical consequence, as suggested in the context of speculative nanoethics (Nordmann 2007). They might be interesting in a merely abstract and philosophical sense to discuss some speculative questions, such as overcoming death, as a kind of thought experiment. There might also be some interest in circles of intellectuals or in the feuilletons of magazines. Yet, regarding the speculative nature of many of those questions, serious concern was expressed that the intellectual effort and the resources spent might be completely irrelevant and wasted (Nordmann and Rip 2009). However, this argumentation is misleading (Grunwald 2010). While techno-visionary futures ranging from high expectations to apocalyptic fears are often more or less fictitious in content, such stories about possible futures can and often do have a real impact on scientific and public discussions (Selin 2007). Even a picture of the future lacking all facticity can influence societal debates, the formation of opinions, issues of perception and acceptance, and even political decision making in two ways at least (following Grunwald 2016):

- Techno-visionary futures can change the way we perceive current and future developments of technology, just as they can change the prospects of future societal constellations. Frequently, the societal and public debate about the opportunities and risks associated with new types of technology revolves around those visionary stories to a considerable extent, as has been the case in the field of nanotechnology (Brune et al. 2006; Fiedeler et al. 2010) and as is still the case in human enhancement (Coenen et al. 2009; Coenen 2010). Visions and expectations motivate and fuel public debate because of the impact the related narratives may hold for everyday life and for the future of important areas of society, such as military, work, and health care. The current debates on future perspectives of digitization are a recent illustration (e.g. Hirsch-Kreinsen 2016). Positive visions can contribute to fascination and public acceptance and also can attract creative young scientists to engage themselves there, just as negative visions and dystopias can cause concern and even mobilize resistance as was feared in particular in the early debate on nanotechnology (Grunwald 2011).

- Techno-visionary futures exert a particularly great influence on the scientific agenda which, as a consequence, partly determines what knowledge will be available and applicable in the future (Dupuy 2007). Directly or indirectly, they influence the views of researchers and, thus, ultimately also exert influence on political support and research funding. For example, the US American funding program on nanotechnology "National Nanotechnology Initiative" (NNI) was named "Shaping the World Atom by Atom" directly referring to visionary ideas of futurist Drexler (1986). Even the speculative stories about improving human performance (Roco and Bainbridge 2002) quickly caused great interest among policy makers and research funders (Nordmann 2004; Coenen et al. 2009). Projections of future developments based on NEST expectations therefore not only might have but really had heavy influence on decisions about the support, funding and prioritization of scientific research and progress. The allocation of research funds is obviously a real intervention which usually will have a real impact on further developments.

In general, the communication of techno-visionary futures represents an *intervention* in ongoing communication. It can trigger a turn in a debate and influence decisions, possibly depending on how consistent, plausible, motivating, threatening or fascinating the respective futures are. The reception of George Orwell's novel *1984* or the consequences of the report of the Club of Rome, *The Limits of Growth*, from 1972 are examples of this. This interventional character can lead to the well-known effects of self-fulfilling or self-destroying prophecies (Merton 1948). Intervening with technology futures in present debates on technology, whether with warnings or hopes, is also a power game linked to values, interests, and intentions (Brown et al. 2000). The factual power of futures in general and techno-visionary futures in particular makes them to an object of responsibility assignments and reflections (Grunwald 2017).

A particular example of this intervention is the story of Drexler's molecular assembler (Drexler 1986). This envisioned machine would be able to form any object by selecting atoms from the environment and positioning them, one at a time, to assemble the object desired: "Eric Drexler (…) believes nano-assemblers could make steaks out of grass, water and foodstuffs, avoiding the cumbersome process involving the cow" (Munich Re 2002, p. 3). In spite of the fact that this highly visionary and even futuristic idea was proven to be not feasible because of natural scientific considerations (Smalley 2001) it exerted a considerable influence on the emergence of nanotechnology as a research field with revolutionary potential, on the emergence of public debate, on motivating contradicting views such as proposed by Bill Joy (2000), on the NBIC movement (Roco and Bainbridge 2002; Wolbring 2008), and on public funding of nanotech research with the National Nanotechnology Initiative (NNI 1999) being the flagship of public funding of nano still today. Thus we can see highly visible consequences and traces of this future proposed more than 30 years ago in spite of the fact that there was no technological advance towards its realization at all.

The factual power of techno-visions thus contributes to shaping the present. Communication and discussion of specific futures can heavily affect a certain present time independent from whether the content of the vision will be realized at some time, independent even from the feasibility of that vision (Lösch 2006). A nice example can be taken from a branch of the German history of nuclear power. In order to close the material flow of radioactive materials in the sense of a circular economy and to realize the vision of a nearly infinite process of energy provision the technology of the Fast Breeder was developed. A reactor of this type was built in Kalkar at the cost of about seven billion *Deutsche Mark*. However, during the construction the acceptance of nuclear power in Germany decreased and almost disappeared, in particular because of the Chernobyl disaster. As a result, the power plant at Kalkar was never taken into operation in spite of the fact that it had been completely finished in 1987. Thus the vision of getting infinite nuclear energy had large real consequences at that time in terms of economic resources to be spent and of social conflict fueled while there has never been a contribution to German energy supply of that technology. The vision contributed to shape the present policies of that time (investments, subsidies by the state, demonstrations of the opponents, growth of the antinuclear movement etc.) while it did not—as was intended—shape the future of the German energy system because it never produced electricity.

3   Reflections: The Immanence of the Present

We make statements about the future, develop pictures of it, simulate temporal developments and create scenarios, formulate expectations and fears, set goals, and consider plans for their realization. All this takes place in the medium of language (Kamlah 1973) and is thus part of the respective *present*. The same holds for statements of the past. Humans always live at a present time while the past and the future are present only in their minds. Epistemologically our standpoint is always the present we are living in and from which we construct our images from the past as well as from the future:

> For if there are times past and future, I wish to know where they are (…) Wherever they are and whatever they are they exist only as present. Although we tell of past things as true, they are drawn out of the memory, not the things themselves, which have already passed, but words constructed from the images of the perceptions which were formed in the mind. (Augustine 397, p. XI, 18)

Also forecasters and techno-visionary writers cannot break out of the present either, always making their predictions on the basis of present knowledge, values, and assessments. Future facts or processes can be neither logically deduced (Goodman 1954) nor empirically investigated. The only things that are empirically accessible are the present images which we make of the future, but not the future presents itself. For this reason, we can talk about possible futures in the plural, about alternative possibilities for imagining the future, and about the justification with which we can expect something in the future. These are always *present futures* and not *future presents* (Luhmann 1992/1998). Therefore, if we talk, for instance, about techno-visionary futures such as cyborgs or far-reaching human enhancement, we are not talking about whether and how these developments will really occur but how we *imagine and assess them today*—and these images and assessment mostly differ to a large extent. Futures are thus something always contemporary and change with the changes in each present (Grunwald 2016).

> When, therefore, they say that future events are seen, it is not the events themselves (…), but perhaps, instead, their causes and their signs are seen, which already do exist. Therefore, to those already beholding these causes and signs, they are not future, but present, and from them future things are predicted because they are conceived in the mind (Augustine 397, p. XI, 18)

Futures do not arise of their own accord. Techno-visionary futures are social constructs. They are man-made and cannot be discovered. Futures are created and disseminated by individual authors, teams, journalists, scientists, and science managers, or they emerge from discourse within scientific communities or at the interface between science and society. Futures, regardless of whether they are forecasts, scenarios, plans, programs, visions, or speculative fears or expectations, are designed using a whole range of ingredients such as available knowledge, value judgments, and suppositions. They are communicated via different channels, journals, networks, mass media, research applications, expert groups,

ELSI or TA projects on policy advice, etc. Some of them, finding no resonance, will quickly disappear within these communication processes while others will "survive" and motivate actors and groups to subscribe to or oppose the visions-in either case the story will continue and the respective futures will have real-world impacts (Selin 2007). Probably, only a few of the visions proposed will find an audience via the mass media and will therefore be able to achieve real impact for public debate and social perception or attitudes at a larger scale. Others may enter the political arena and result in political decisions, e.g. about research funding (see above). Also the significance of visionary thinking in specific NEST fields may vary over time (Lösch 2010). The history of spaceflight, for instance, is full of techno-visionary promises which regularly fail but nevertheless survive and remain fascinating to many people. The narratives of human settlements on the Mars or on artificial space stations belong to those persistent stories which again and again have real-world impacts by attracting a lot of research funding.

Because futures are man-made, they have authors involving intentions, objectives and purposes. The designing of futures is purposive action, intended for example to provide orientation, to create fascination, to promote a certain line of development, to attract research funding, to raise awareness in the public, to initiate a debate, or to support partisan interests. By framing the construction of futures in the means/end rationality questions of the following type come into consideration (Grunwald 2016): Which actors—individuals as well as collectives such as project groups, institutes, or associations—belong to the authors? Which perspectives do they include and express? Which motives are they pursuing? What ideas do they have about the relationship between technology and society? What is their stand in general toward scientific, technological progress? To which contexts, networks, policy groups, pressure groups, etc. can they be assigned? Why and for what purpose was a specific techno-future designed? What shall a proposal for a definition bring about? Which diagnoses, values, or even interests are behind this choice of aims? Are there different and perhaps conflicting goals and purposes pursued by different actors? Hermeneutic analysis and discourse analysis can shed light on this unordered list of questions and help enlightening the background of the futures created and communicated.

In the construction of techno-visionary futures numerous decisions must be made about the purposes pursued (see above) and the means identified as appropriate to reach the purposes. In particular, building techno-visionary futures needs ingredients such as background data, knowledge about regularities or correlations, assumptions and estimates of relevance as well as a process of composing these ingredients into a coherent picture of the future. In order to better understand these futures in content and with regard to their background motivation the question has to be answered as to what ingredients are used in the shaping of futures and in which way these ingredients have been assembled and composed in arriving at the respective statements about the future. As far as their knowledge structure is concerned, futures are initially opaque constructs consisting of highly diverse elements. In a rough approximation, the following gradation of knowledge and other components can initially be made without claiming completeness (Grunwald 2016):

- *Present knowledge* which is proven according to accepted criteria (e.g., of the respective scientific disciplines) to be knowledge (e.g., according to the issue at stake from the field of nanotechnology, engineering, economics);
- *Estimates* of future developments that do not represent current knowledge but that can be substantiated by current knowledge (e.g., demographic change, energy needs, velocity of the technological advance);
- *Values* and normative expectations about the future society, future relations between humans and technology, or between society and nature, etc.;
- *Ceteris paribus (all other things being equal) conditions*, whereby certain continuities—business as usual in some sense or a lack of disruptive changes—can be assumed as a framework for the prospective statements;
- Ad hoc *suppositions* which are not substantiated by knowledge, but taken as given (e.g., the future validity of a German phase-out of nuclear energy, or the non-occurrence of a catastrophic impact of a comet on the Earth);
- *Utopian ideas* of worlds where everything could be different in the future, speculative proposals for futures worlds, science fiction stories and other imaginations.

Futures are thus created in accordance with available knowledge, but also with references to assessments of relevance, value judgments, and interests, perhaps normative visions and utopias and also may include mere speculation. The construct character of futures can thus be exploited, on the one hand, by those representing specific positions on social issues, substantial values, and particular interests in order to produce future visions corresponding to their interests and to employ these to assert their particular positions in debates (Brown et al. 2000). On the other, this construct character calls for enlightening analysis of ingredients and composition in order to provide better understanding.

These considerations are in accordance with the reflections on time given by Augustine of Hippo (see above). Futures are created in the present based on experiences from the past and expectations of the future which, however, both are also constructs made in present time. We cannot escape the immanence of the present. Therefore, our interventions cannot affect the future present directly but the respective present time only.

4 Consequences for Technology Assessment

Technology assessment (TA) constitutes a research-based response to challenges and problems at the interface between science and technology on the one hand, and humans and society, on the other (Grunwald 2010, 2019). Its specific mission is creating *knowledge for action* in shaping the technological advance and the usage of its outcomes. TA consists of a combination of *knowledge production* concerning possible consequences of new technology, the transparent *evaluation* of this knowledge from a societal perspective involving values and goals, the *development of options* how to proceed, and of making knowledge and options available to politics and society. TA is both interdisciplinary and transdisciplinary in nature. Recently the following definition of TA was proposed (Grunwald 2019):

> Technology assessment is a set of socio-epistemic practices[2] serving the cognitive interest of enhancing reflectivity for shaping the technological advance and the usage of its outcomes in a democratic way. Enhancing reflectivity shall be realized by providing and assessing prospective knowledge, by applying an inclusive approach to a diversity of perspectives in social and epistemic respect, and by applying systems thinking.

TA arose from specific historical circumstances in the 1960s and 1970s. Concerns in the U.S. political system, in particular in the Congress, culminated in the creation of the Office of Technology Assessment (OTA) in 1972 (Bimber 1996) as the first manifestation of TA. Parallel to this specific development in the political system, radical intellectual changes took place and resulted in the more general motivations of TA (Grunwald 2019). TA was then expected to contribute to new forms of societal orientation and legitimisation of science and technology facing those

---

[2]The notion of socio-epistemic practices refers to a twofold constellation characteristic to TA: (1) its social processes of involving different actor groups are not only relevant in social but also in epistemic respect; and (2) the results of these socio-epistemic processes include knowledge for action which is not only an epistemic object but also may restructure social conditions how to act and decide.

challenges: to explore possible unintended and negative side-effects of technology (Bechmann et al. 2007), to assess and weigh risks and chances, to reconcile technology and democracy, to elaborate strategies for legitimate decision-making, to help resolving technology conflicts. The first phase of TA was characterized by an underlying belief in technology determinism (Ropohl 1982) and planning optimism. TA was expected to provide predictions of future technologies and their consequences in order to allow society and politics to better adapt to those consequences.

Since the 1980s the idea of technology determinism was overcome in favour of the opportunity of *shaping technology* by early reflection on possible later impacts and consequences (Bijker and Law 1994). The adaptation of the social constructivist programme (Bijker et al. 1987) to TA was performed within the approach of Constructive Technology Assessment (CTA). CTA should and still shall serve as a reflexive think tank in society with the mission to contribute to "a better technology in a better society" (Rip et al. 1995). Major current objectives of TA are providing contributions to shaping technology towards a more sustainable development as well as to the RRI movement (Owen et al. 2013; van den Hoven et al. 2014).

In each of the concepts proposed TA explores and assesses, on the one hand, possible impacts and consequences of technology in a prospective manner. On the other, it helps to introduce society's expectations and needs concerning new technology into the agenda-setting processes for research and development. Regarding that there are influences in both directions, from technology development on society, but also from society's expectations on technology, the notion of a "co-evolution" of technology and society gained much acceptance (Rip 2007), with technology assessment acting as its medium.

Anyway, the TA focus on future consequences of technology, be they intended or unintended, puts it under the consequentialist paradigm (Fig. 1). In the German translation *Technikfolgenabschätzung* even the notion of consequences (*Folgen*) has been made part of the term. TA investigates consequences which could perhaps, plausibly or probably become reality in the futures. Prospective knowledge in TA is knowledge about consequences which do not yet exist and perhaps will never become reality. Providing prospective knowledge has a merely instrumental function in TA and is not an end in itself: the prospective knowledge is the object of assessing, reflecting, evaluating and judging, e.g. with respect to desirability, acceptability, or responsibility. At the end of this consequentialist reasoning conclusions for action and decision-making are drawn based on provision and assessment of the prospective knowledge. This consequentialist loop starts in a

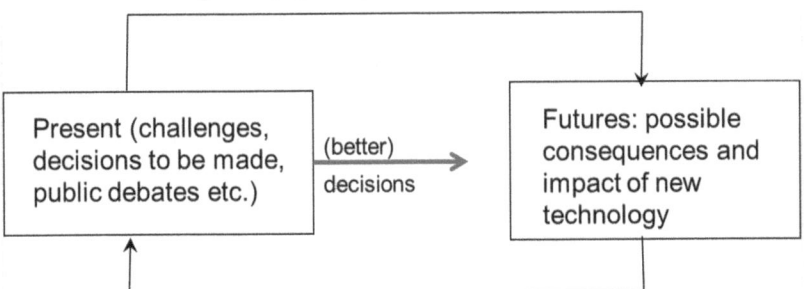

Creation of pictures of the future for alternative options: predictions, scenarios, expectations, fears, visions and so forth as diagrams, figures, narratives etc.

Present (challenges, decisions to be made, public debates etc.) (better) decisions → Futures: possible consequences and impact of new technology

Assessment and reasoning: comparison of the alternatives under consideration, evaluation with respect to desirability, drawing consequences for decisions to be made, developing action strategies, measures, etc.

Fig. 1 The consequentialist loop of technology assessment

respective present, leads to considering some futures, and then leads back to the respective present (Fig. 1).

According to the analysis given in the sections following Augustine above two immediate conclusions can be drawn:

1. The diagnosis of the immanence of the present also applies to technology assessment which means that the consequentialist loop (Fig. 1) entirely takes place in a respective present—the futures considered are merely present constructions and images of futures, and
2. Decisions made upon advice of TA do not shape the future but rather the respective present. They are interventions into the respective present time. Whether these interventions will lead to the intended outcomes and consequences remains an expectation. The future development only will show whether they will be fulfilled or to which extent they will be fulfilled, or which unintended effects also might emerge.

While the first conclusion seems to be common sense in TA the second is not. At least many TA practitioners (including the author) again and again speak of the mission of TA as "contributing to shaping the future". However, taking conclusion (2) seriously implies that TA should no longer be conceptualized as supporting shaping socio-technical futures but rather as helping shaping the present according to pictures of the future created, debated, and assessed. By doing this the familiar claim of shaping the future in the usual sense of a future present does not completely disappear. It only moves more to the background because in the time span between the respective intervention into a specific present time and the intended shaping of future events or developments a lot of things may happen which had not been taken into account and which could make the relation between the intervention and the consequences after some time much looser than intended or expected.

This shift of consideration might perhaps appear artificial, too sophisticated, and perhaps without any relevance to TA's daily business. In the remainder of this section, however, I will demonstrate that there are relevant lessons to be learned from this shift, at least in certain TA constellations.

4.1   Awareness-Raising by Trend Extrapolation

According to the considerations given in Sect. 3 in accordance with Augustine's view on time it is obvious that extrapolations do not provide any knowledge of future presents. Rather they but are expressions of *current* respective *past* knowledge which is applied to future developments under the presupposition that things simply will go on. Extrapolations, in particular trend extrapolations are the mere continuation of the past to the future in the present. With Augustine we might say that trend extrapolations identify the knowledge about past developments with expectations concerning the future altogether in present time. While they therefore do not tell any truth about the future, they can be used to shed some light on (perhaps alarming) *current* developments and have a merely instrumental function. This was indeed the most important effect of famous future stories: when, for example, Rachel Carson postulated the "Silent Spring" in 1962 she envisaged a future world without birds which suffered progressively from DDT-pesticides in those days. This projection however did not realize in the decades after because she underestimated the impact of her report on society, which effectively banned the pesticide for the sake of the environment, subsequently.[3] However, there is no

---

[3]This is a nice example for the interventionist power of futures for the respective present (Merton 1948, cp. Sect. 3).

doubt that Carson took up a serious observation of her time. By extrapolating it to the future she created awareness what might happen if no countermeasures would be taken. In this sense, trend extrapolations should less pretend to know the future as they could better give orientation for action what should be done in shaping the present (van der Burg 2014). The story of the "Limits to Growth" published by the Club of Rome in 1972 showed a similar effect. Pictures of the future based on the extrapolation of trends often help to raise awareness concerning problematic developments. This type of futures is as an illustrative example for futures shaping the present—which is an important mechanism of gaining orientation and prioritizing activities. And it is an example of the immanence of the present because it does not tell any certain message about a future present.

4.2 Hermeneutic Extension of Technology Assessment

The observation that also the consequentialist loop remains trapped in the immanence of the present together with the situation of little or almost no valid prospective knowledge available virulent in many NEST fields gave rise to propose an hermeneutic extension of the traditional consequentialist approach of TA (Grunwald 2014, 2016). The hermeneutic analysis of pictures of the future hardly tells us anything about the future in the sense of a present in the time to come, but rather *about us today*. If projections of the future are interpreted in a way that makes it clear why we aggregate certain current ingredients to specific futures and argue dedicatedly about them, then we have learned something *explicitly* about ourselves, our societal practices, subliminal concerns, implicit hopes and fears, and their cultural roots. Compared to the consequentialist paradigm with its central focus on questions as to the possible impacts of new technologies, how we assess these, and whether and under what circumstances we welcome or reject these implications, this perspective includes a set of questions into TA reasoning which were not that significant before (following Grunwald 2014):

- What are the implications of the new developments in science and technology for the present and future of man and society, which fundamental constellations (man/technology, man/nature, etc.) do they change, and "what is at stake?" e.g. in ethical, cultural, and social terms?
- How is a philosophical, ethical, social, cultural, etc. significance attributed to scientific-technological developments, which after all are nothing initially but scientific-technological developments? What role do e.g. (visionary) techno futures play in this context?

- How are attributions of meaning being communicated and discussed? What roles do they play in the major technological debates of our time? What forms of communication and linguistic resources are being used and why? What extra-linguistic resources (e.g. movies, works of art) play a role in this context and what does their use reveal?
- Why do we discuss scientific-technological developments in the way we do and with the respective attributions of meaning rather than in some other way?
- How does man as an historical being see himself in discourses about techno-futures? What future concepts are being applied if the future is presented either as though it were possible to shape it technically or politically, or as what will contingently come about and will never be quite adequate in terms of a historical responsibility to bring about a better world?

Responding to these question needs extended interdisciplinary work including contributions from sociological and philosophical discourse analysis, linguistics, the historical science, and cultural studies.

4.3 Democratic Dimension: Opening up Futures Instead of Closure

The emphasis on the open space of futures in hermeneutic orientation is cognate with 'opening up' the outputs of TA with plural and conditional advice (Ely et al. 2014). While frequently scientific reasoning about futures is in risk of closing them down to a presumed "one best solution" in the attitude that "science knows best" the analysis given above about the immanence of the present (Stirling 2008) also holds for scientifically produced futures. Keeping alternative futures open or opening up new ones by elaborating on further alternative options in TA's tradition of "thinking in alternatives" (Grunwald 2019), however comes at a price. While many actors in politics and public debate claim that complexity must be reduced and that things must be made as simple as possible TA usually does the opposite. Instead of reducing complexity it adds further future options involving consequences and impacts of new technology to the processes of reasoning and weighing. The obligation of TA to be reflective renders simplistic views on new technology or transformations processes as well as the closure of futures impossible. This implies that more time, resources, and effort will be needed for reasoning, deliberation, and decision-making, and that TA is in charge of developing procedures to deal with this increased complexity.

5   Concluding Remark

Technology assessment and other research activities investigating and reflecting the scientific-technological advance and the usage of its outcomes have become more modest concerning the possibility of predicting future technologies and their impacts on society. In particular, it is the issue of the 'co-evolution' of technology and society which strongly limits today's trends and assumptions to the future. Notions such as techno-visionary futures and present futures using the term 'future' in the plural illustrate this development.

While this modesty seems to be common-sense in many communities today (except the newly emerging optimism of enabling better prognoses by Big Data technologies) the phrase about shaping the future still can be observed everywhere in engineering, in the economy, in the public and in politics. In this chapter I have proved that this notion is misleading in its frequent naïve usage. Therefore, a more reflected view on this and related terms should be applied.

References

Augustine of Hippo (397). *Confessions* XI, 20.

Bechmann, G., Decker, M., Fiedeler, U., & Krings, B. (2007). TA in a complex world. *International Journal of Foresight and Innovation Policy, 4,* 4–21.

Bijker, W. E., & Law, J. (Eds.). (1994). *Shaping technology/building society.* Cambridge: MIT Press.

Bijker, W. E., Hughes, T. P., & Pinch, T. J. (Eds.). (1987). *The social construction of technological systems. New directions in the sociology and history of technological systems.* Cambridge: MIT Press.

Bimber, B. A. (1996). *The politics of expertise in Congress: The rise and fall of the office of technology assessment.* New York: State University of New York Press.

Brown, J., Rappert, B., & Webster, A. (Eds.). (2000). *Contested futures. A sociology of prospective techno-science.* Burlington: Ashgate.

Brune, H., Ernst, H., Grunwald, A., Grünwald, W., Hofmann, H., Krug, H., et al. (2006). *Nanotechnology-Perspectives and assessment.* Berlin: Springer.

Camhis, M. (1979). *Planning theory and philosophy.* London: Law Book Co of Australasia.

Coenen, C. (2010). Deliberating visions: The case of human enhancement in the discourse on nanotechnology and convergence. In M. Kaiser, M. Kurath, S. Maasen, & C. Rehmann-Sutter (Eds.), *Governing future technologies. Nanotechnology and the rise of an assessment regime* (pp. 73–87). Dordrecht: Springer.

Coenen, C., Schuijff, M., Smits, M., Klaassen, P., Hennen, L., Rader, M., & Wolbring, G. (2009). Human enhancement. Brussels: European Parliament. http://www.itas.fzk. de/deu/lit/2009/coua09a.pdf.

Collingridge, D. (1980). *The social control of technology.* London: Pinter.

Drexler, K. E. (1986). *Engines of creation-The coming era of nanotechnology.* Oxford: Anchor Books.

Dupuy, J.-P. (2007). Complexity and uncertainty: A prudential approach to nanotechnology. In F. Allhoff, P. Lin, J. Moor, & J. Weckert (Eds.), *Nanoethics. The ethical and social implications of nanotechnology* (pp. 119–132). Hoboken: Wiley.

Ely, A., Van Zwanenberg, P., & Stirling, A. (2014). Broadening out and opening up technology assessment: Approaches to enhance international development, co-ordination and democratisation. *Research Policy, 43*(3), 505–518.

Fiedeler, U., Coenen, C., Davies, S. R., & Ferrari, A. (Eds.). (2010). *Understanding nanotechnology: Philosophy, policy and publics.* Heidelberg: AKA.

Giese, B., Pade, C., Wigger, H., & von Gleich, A. (Eds.). (2014). *Synthetic biology: Character and impact.* Heidelberg: Springer.

Goodman, N. (1954). *Fact fiction forecast.* Cambridge: Harvard University Press.

Grin, J., & Grunwald, A. (Eds.). (2000). *Vision assessment: Shaping technology in 21st century society.* Berlin: Springer.

Grunwald, A. (2006). Nanotechnologie als Chiffre der Zukunft. In A. Nordmann, J. Schummer, & A. Schwarz (Eds.), *Nanotechnologien im Kontext* (pp. 49–80). Berlin: AKA.

Grunwald, A. (2007). Converging technologies: Visions, increased contingencies of the conditio humana, and search for orientation. *Futures, 39*(4), 380–392.

Grunwald, A. (2010). From speculative nanoethics to explorative philosophy of nanotechnology. *NanoEthics, 4*(2), 91–101.

Grunwald, A. (2011). Ten years of research on nanotechnology and society-Outcomes and achievements. In T. B. Zülsdorf, C. Coenen, A. Ferrari, U. Fiedeler, C. Milburn, & M. Wienroth (Eds.), *Quantum engagements: Social reflections of nanoscience and emerging technologies. Proceedings der S.Net Konferenz 2010 in Darmstadt* (pp. 41–58). Heidelberg: AKA.

Grunwald, A. (2013). Techno-visionary sciences: Challenges to policy advice. *Science, Technology & Innovation Studies, 9*(2), 21–38.

Grunwald, A. (2014). The hermeneutic side of responsible research and innovation. *Journal of Responsible Innovation, 1*(3), 274–291.

Grunwald, A. (2016). *The hermeneutic side of responsible research and innovation.* London: Wiley-ISTE.

Grunwald, A. (2017). Assigning meaning to NEST by technology futures: extended responsibility of technology assessment in RRI. *Journal of Responsible Innovation, 4,* 100–117.

Grunwald, A. (2019). *Technology assessment in practice and theory.* Oxford: Routledge.

Hirsch-Kreinsen, H. (2016). *"Industry 4.0" as promising technology: Emergence, semantics and ambivalent character.* Soziologisches Arbeitspapier, Nr. 48/2016. Dortmund: Universität Dortmund.

Joy, B. (2000 Apr.). Why the future does not need us. *Wired Magazine,* 238–263.

Kamlah, W. (1973). *Philosophische Anthropologie: Sprachkritische Grundlegung und Ethik.* Mannheim: Bibliographisches Institut.

Lösch, A. (2006). Means of communicating innovations. A case study for the analysis and assessment of nanotechnology's futuristic visions. *Science, Technology & Innovation Studies, 2,* 103–125.

Lösch, A. (2010). Visual dynamics: The defuturization of the popular 'Nano-Discourse' as an effect of increasing economization. In M. Kaiser, M. Kurath, S. Maasen, & C. Rehmann-Sutter (Eds.), *Governing future technologies. Nanotechnology and the rise of an assessment regime* (pp. 89–108). Dordrecht: Springer.

Luhmann, N. (1998). *Describing the future. Observations on Modernity* (trans: Whobrey W.). Stanford: Stanford University Press (First publication 1992).

Merton, R. (1948). The self-fulfilling prophecy. *The Antioch Review, 8*(2), 193–210.

Munich Re. (2002). Nanotechnology-What is in store for us? http://www.munichre.com/publications/302-03534_en.pdf.

NNI – National Nanotechnology Initiative. (1999). National Nanotechnology Initiative. Washington.

Nordmann, A. (2004). *Converging technologies-Shaping the future of European societies.* In high level expert group "Foresighting the new technology wave". Luxembourg: Office for Official Publications of the European Communities.

Nordmann, A. (2007). If and then: A critique of speculative nanoethics. *Nanoethics, 1,* 31–46.

Nordmann, A., & Rip, A. (2009). Mind the gap revisited. *Nature Nanotechnology, 4,* 273–274.

Owen, R., Bessant, J., & Heintz, M. (Eds.). (2013). *Responsible innovation: Managing the responsible emergence of science and innovation in society.* Hoboken: Wiley.

Rip, A. (2007). Die Verzahnung von technologischen und sozialen Determinismen und die Ambivalenzen von Handlungsträgerschaft im 'Constructive Technology Assessment'. In U. Dolata & R. Werle (Eds.), *Gesellschaft und die Macht der Technik. Sozioökonomischer und institutioneller Wandel durch Technisierung* (pp. 83–106). Frankfurt a. M.: Campus.

Rip, A., Misa, T. J., & Schot, J. (Eds.). (1995). *Managing technology in society: The approach of constructive technology assessment.* London: Pinter.

Roco, M. C., & Bainbridge, W. S. (Eds.). (2002). *Converging technologies for improving human performance.* Arlington: Kluwer Academic.

Ropohl, G. (1982). Kritik des technologischen Determinismus. In F. Rapp & P. T. Durbin (Eds.), *Technikphilosophie in der Diskussion* (p. 3.18). Braunschweig: Vieweg.

Selin, C. (2007). Expectations and the emergence of nanotechnology. *Science, Technology and Human Values, 32*(2), 196–220.

Smalley, R. E. (2001). Of chemistry, love and nanobots. *Scientific American, 285,* 76–77.

Stirling, A. (2008). "Opening up" and "closing down": Power, participation, and pluralism in the social appraisal of technology. *Science, Technology and Human Values, 33*(2), 262–294.

Van den Hoven, J., Doorn, N., Swierstra, T., Koops, B.-J., & Romijn, H. (Eds.). (2014). *Responsible innovation 1: Innovative solutions for global issues.* Dordrecht: Springer.

van der Burg, S. (2014). On the hermeneutic need for future anticipation. *Journal of Responsible Innovation, 1*(1), 99–102.

Voss, J.-P., Bauknecht, D., & Kemp, R. (Eds.). (2006). *Reflexive governance for sustainable development.* Cheltenham: Edward Elgar.

Wolbring, G. (2008). Why NBIC? Why human performance enhancement? *The European Journal of Social Science Research, 21,* 25–40.

Armin Grunwald (Ph.D. and habilitation) is Director of the Institute for Technology Assessment and Systems Analysis and Full Professor of Philosophy and Ethics of Technology at the Karlsruhe Institute of Technology. Simultaneously, he heads the Office of Technology Assessment at the German Bundestag.

Prototype Scenarios as Negotiation Arenas Between the Present and Imagined Futures. Representation and Negotiation Power in Constructing New Socio-Technical Configurations

Ingo Schulz-Schaeffer and Martin Meister

Abstract

Situational scenarios and prototype scenarios in particular are socio-technical futures that shape present socio-technical developments in a specific way. Scenario-building is about putting together heterogeneous components, some of which are already existing and some of which still have to be created. The requirement to adapt these present and future components to each other turns scenario-building into negotiating between the present and imagined futures. In this contribution, we focus on how the envisaged contexts of use are represented in these negotiations. Based on empirical research on technology development in the field of ubiquitous engineering, we distinguish between different forms of representation and assess their respective negotiation power.

I. Schulz-Schaeffer (✉) · M. Meister
Department of Sociology, TU Berlin, Berlin, Germany
e-mail: schulz-schaeffer@tu-berlin.de

M. Meister
e-mail: martin.meister@tu-berlin.de

© Springer Fachmedien Wiesbaden GmbH, part of Springer Nature 2019
A. Lösch et al. (eds.), *Socio-Technical Futures Shaping the Present*,
Technikzukünfte, Wissenschaft und Gesellschaft / Futures of Technology,
Science and Society, https://doi.org/10.1007/978-3-658-27155-8_3

1   Introduction

Socio-technical futures shape the present and in doing so they shape the future. The term "socio-technical futures" was coined to emphasize the fact that our current notions about future technology and their future uses do not have the status of more or less accurate predictions but rather of imaginations whose relevance largely lies in how they affect current thoughts, debates and actions (cf. Grunwald 2009, 2012; Grin and Grunwald 2000). From this point of view, there is not one socio-technical future but a multitude of possible socio-technical futures. If one of those futures eventually becomes reality, it is not at least because the respective visions, scenarios, or roadmaps attracted people, mobilized resources, influenced societal discourses, and guided innovation-related activities (cf. Borup et al. 2006; van Lente and Rip 1998; Schulz-Schaeffer and Meister 2015). Because of this "performative dimension of future images" (Geels and Smit 2000, p. 882; van Lente 2012, p. 772) they shape the future by shaping the present. Furthermore, the term socio-technical futures allows to take into account that there are quite different forms of future images such as national sociotechnical imaginaries (cf. Jasanoff and Kim 2009), visions (cf. Sturken et al. 2004), guiding visions (cf. Dierkes et al. 1992), techno-visionary futures (cf. Grunwald 2013), scenarios (cf. Kahn and Wiener 1967; Schulz-Schaeffer 2013), or roadmaps (cf. de Laat 2004).

These forms of future images influence current thoughts, debates and actions in different ways and, thus, contribute in different ways to creating the future. Roadmaps, for instance, have a tendency to turn into self-fulfilling prophecies (cf. Schaller 1997). By precisely specifying assumed future technological requirements and the steps to be taken in order to meet those, roadmaps can (and often are intended to) trigger a dynamic where the innovating actors make the assumptions come true by acting according to these specifications (cf. Sydow et al. 2005). Visions, on the other hand, are influential mainly because of their capacity to evoke expectations about promising futures, thus attracting attention and mobilizing actors and resources (cf. Borup et al. 2006; van Lente and Rip 1998; van Lente 1993). In this paper, we will focus on situational scenarios, yet another form of future images. Situational scenarios are future images that in some detail specify for envisaged typical situations of use of an imagined new technology how the components of these situations would (or might) interact (cf. Schulz-Schaeffer and Meister 2015, 2017). Situational scenarios create possible socio-technical futures in a nutshell, which then can be used and are used by technology developers or other stakeholders as guidance for actually realizing (or for preventing) this future.

The contributors to this edited volume agree that current engagements with possible socio-technical futures always involve a two-way process of negotiation, including negotiation about how the present ought to be transformed towards the imagined future as well as about how the imagined future would have to take into account current states. In this paper, we examine the process of negotiation between the present and possible futures as mediated through situational scenarios and especially by situational scenarios that are prototypically realized. We will argue that situational scenarios are combinations of components from current situations and of components that will come into existence only as part of the imagined future. The former we will call "present components", the latter "future components". Scenario-building means to interrelate the scenario's components by adapting them to each other. Thus, situational scenarios contain decisions about which components would induce others to change and which components would have to be modified (or exchanged), if this scenario became reality. For this reason we suggest to view scenario-building as a negotiation process (cf. Schulz-Schaeffer and Meister 2017). It also is a process of negotiation between the present and possible futures since it is about whether future components will have to adapt to present components or vice versa. To elaborate on how situational scenarios accommodate negotiations between present and future components, we will adopt Anselm Strauss' concept of arena.

The considerations of this paper build on about 60 interviews with researchers from the field of ubiquitous computing, which we conducted 2013 in the USA, in Japan, and in the EU. Our findings rely on grounded theory-based interpretations of the transcribed interviews combined with content analyses of scientific publications and related research documents. We gathered the material as part of the research project "Scenarios as patterns of orientation in technology development and technology assessment" (funded by the German Research Foundation, DFG).

The remainder of this paper is structured as follows. In Sect. 2, we describe our concept of situational scenario and elaborate on the distinction between present and future components. In Sect. 3, we explain why we conceive scenario-building as a process of negotiation between the present and imagined futures. In Sect. 4, we introduce the perspective of viewing situational scenarios as negotiation arenas. In Sect. 5, we apply this perspective to prototype scenarios and ask who represents the (present and future) components in negotiations. For reasons that will become clear later in the article, we focus on different forms of representation of the components related to the envisaged context of application of the imagined new technology. In Sect. 6, we provide empirical examples for these different forms of representation and evaluate their respective negotiation power. In the concluding section, we briefly summarize our conceptual considerations and empirical findings.

2   Situational Scenarios and Their Present and Future Components

According to our definition, situational scenarios are images of the future that in some detail specify for envisaged typical situations of use of an imagined new technology how the components of these situations would (or might) interact. These components include not only the imagined new technology with its features and the envisaged users with their interests, preferences and capabilities, but also other people, objects or structures of relevance for the situation. Situational scenarios provide descriptions of the interactions between the components included. They focus attention to the causal relationships one would have to take into account, if the scenario was reality (cf. Schulz-Schaeffer and Meister 2015, 2017, p. 166, 2). In the work of the research engineers in the field of ubiquitous computing we studied, situational scenarios occur in different manifestations. Often they are represented as narrative scenarios, that is, as symbolic representations of the respective scenario as written story, video, or comic. Sometimes they remain unarticulated, existing as implicit scenarios, as tacit mental images, and thus observable only by their effects on the engineers' work. In the engineers' laboratories situational scenarios often become prototypically realized, thus attaining a form we call prototype scenarios. Prototype scenarios, in our definition, are more or less comprehensive physical implementations of the components that make up the underlying narrative (or implicit) situational scenario within more or less realistic laboratory settings (cf. Schulz-Schaeffer and Meister 2017, p. 5, 8–11). Situational scenarios represent a kind of imagined socio-technical future that is influential when it comes to informing concrete conceptual and design decisions of technology developers. Our previous empirical studies on technology development in the field of ubiquitous computing provided evidence for this claim.

Situational scenarios that are employed in technology development represent always a combination of components from current situations and of components that will come into existence only as part of the imagined future. The former we will call "present components", the latter "future components". Since situational scenarios are images of the future, it is obvious that they should include future components. For example, the new technology to be developed is a future component, but far from being the only one. Within situational scenarios, especially when they are prototypical realized, the future components are "already here".[1]

---

[1]In the sense of the well-known quote attributed to William Gibson 'The future is already here—it's just not evenly distributed'.

They exist as part of an imagined future reality that has become concrete enough to be spelled out as a consistent narrative or even as a physical arrangement within a laboratory.

To come up with ideas about technological innovations is an activity that takes place in the present and thus rests upon today's knowledge, believes and assumptions. Thus, when engineers are searching for new technological solutions for problems or when they aim at technologies with desirable new features, the problems or desires they have in mind are usually derived from current problems and desires. Consequently, their images of the future rely on explicit or implicit assumptions about similarities between the imagined future situations and corresponding present situations. This is especially true with respect to situational scenarios that are employed in actual technology development. Because these scenarios are not about some distant future, but rather envision innovations that engineers hope to realize—at least prototypically—within the timeframe of their research and development projects. For these reasons, a considerable number of the components of situational scenarios are projections of aspects of current situations. These are what we call the scenario's present components. We will further elaborate on our understanding of present and future components below in Sect. 5.

3   Scenario-Building as Negotiation Process

Building, specifying and evaluating situational scenarios means to interrelate their components in particular ways. We suggest to view these processes as negotiation processes (cf. Schulz-Schaeffer and Meister 2017). As mentioned above, situational scenarios are descriptions of the causal relationships between the components of an imagined future situation. The relationships between the components as described in these scenarios require the components to be adapted to one another accordingly. Thus constructing situational scenarios means to interrelate components by adapting them to each other. For instance, if one includes a particular future component in a situational scenario or assumes that a particular present component will stay the same this has implications with respect to the characteristics of other components of the scenario. Defining one component in a particular way requires changing other components accordingly, in order to align their activities and to make sure that the scenario as a whole provides the envisioned features.

This process of adaption between the components in situational scenarios is similar to the process of network building between heterogeneous entities as described and analyzed by actor-network theory. According to actor-network theory, the entities that eventually make up an actor-network are at the same time cause of change and subject to change. On the one hand, they are involved in attempts to define or redefine other components of the emerging network. On the other hand, they are themselves subject to such attempts (cf. Latour 1991, 2005, p. 109, 124, 52–53; Schulz-Schaeffer 2014, 2017). These (re-)definitions are necessarily connected to each other: Successful (re-)definitions of entities (which are called inscriptions in actor-network theory), are accompanied by prescriptions, that is requests to other entities to change their behavior accordingly (cf. Latour 1988, p. 306). As Michel Callon has pointed out, to the degree the activities of the entities involved fit together despite their heterogeneity, they build convergent actor-networks (cf. Callon 1991, p. 148). Mutual definition and redefinition of components is the process that facilitates the emergence of convergent networks. In the same way, it requires mutual adaption between the components in order to obtain situational scenarios that display internally consistent descriptions of possible futures.

Introducing a new component or changing a component's characteristics requires changing other components' characteristics in order to maintain (or even better: to enhance) a scenario's internal consistency. Thus, situational scenarios contain decisions about which components would induce others to change and which components would have to be modified (or exchanged), if this scenario became reality. These decisions have implications for which of the parties involved would profit from these changes in the imagined future and which parties would have to adapt to them. For this reason, we believe it necessary to conceptualize the process of building, specifying and evaluating situational scenarios by which the scenario's components and their characteristics take shape as a negotiation process. It also is a process of negotiation between the present and possible futures since building, specifying and evaluating situational scenarios is always about whether future components will have to adapt to present components or vice versa.

4   Applying the Concept of Arena to Situational Scenarios

To elaborate on how situational scenarios accommodate negotiations between present and future components, we conceptualize them as arenas of negotiation between the present and imagined futures. Following Anselm Strauss, by "arena" we mean the social settings where „disagreement about directions of action"

(Strauss 1993, pp. 226–227) is negotiated. The term does primarily refer to symbolic spaces, that is, to particular meaningful frames of reference that facilitate processes of negotiation. Often, however, these symbolic spaces are accompanied by particular physical spaces where (some of) the negotiations actually take place: the parliament in the case of policy arenas or the academic conference in the case of science arenas (cf. Strauss 2016 [1991]).

As Strauss emphasizes, arenas are spaces of negotiation between different modes of perception and interpretation shared by the members of different social worlds[2] rather than between individuals: "The concept of arena will refer here to interaction by social worlds around issues-where actions concerning these are debated, fought out, negotiated, manipulated, and even coerced within and among the social worlds. It can be individuals who do the acting, but for sociological purposes we want to locate them in some sort of social unit." (Strauss 1993, p. 226) Though, even if individuals debate and negotiate the controversial issues, these debates and negotiations are not about their individual points of view. Rather, they act as representatives of the social worlds or sub-worlds involved, that is, as representatives and spokespersons of the different complexes of meaning of these social worlds.

Processes of innovation are processes of negotiation between different social worlds. They are processes in which quite different complexes of meaning come into in play, generating differences in how things are perceived and interpreted, differences that have to be negotiated in order to proceed. Among them are the particular perceptions and views shared by the engineering disciplines or engineering specialties involved, and the perceptions and views within the envisaged contexts of application of the technology to be developed. In addition, there are the perceptions and views held by the economic, political and legal actors, which are involved in organizing, financing, and regulating the innovation process.

In the remainder of this paper, we will focus on the social worlds and the associated complexes of meaning of the engineering disciplines (or specialties) and of the envisaged contexts of application. Since situational scenarios are descriptions

---

[2]Citing Adele Clark (1991, p. 131); Strauss (1993, p. 212) defines social worlds as "groups with shared commitments to certain activities, sharing resources of many kinds to achieve their goals, and building shared ideologies about how to go about their business". Additional features according to Strauss are that in "each social world, at least one primary activity [...] is strikingly evident", that there are "sites where activities occur", and that most worlds "evolve quite complex technologies" and organizations that further the social world's activities (Strauss 1993, p. 212–213).

of the future application and use of future technology, they are predominantly influenced by ideas and notions related to these two social contexts. This is even more true with respect to the subject of our empirical research: situational scenarios built and used by research engineers as part of their work in academic research laboratories. Innovation work in these settings primarily is about inventing new technological principles, procedures, and solutions and demonstrating that they in principle work as expected within their envisaged contexts of application. For this reason, organizational, financial, and regulatory issues that are important for turning innovations into marketable products do not play a major part in these scenarios.

A situational scenario's plot can provide a quite specific frame of reference for the different ways the scenario's components could interrelate. Situational scenarios, therefore, can serve as especially clear-cut arenas of negotiation. In addition, they have the effect of increasing the pressure to reach an agreement, because the failure to adapt the components to each other puts the whole scenario at risk. As we argue below, this applies even more to prototype scenarios.

5   Prototype Scenarios as Negotiation Arenas

We define prototype scenarios as more or less comprehensive physical implementations of the components that make up the underlying narrative (or implicit) situational scenario within more or less realistic laboratory settings. Prototype scenarios are situational scenarios that are materially realized. "Materially realized" means that there are technical prototypes representing the main technological components of the innovation. Additionally, there is a testbed that includes the most relevant other components of the scenario in a manifestation that allows for real (and not only imagined) interactions between and among the technological and these other components. These other components usually cover the most relevant aspects of the envisaged context of application. Typically, they include the intended users with their characteristics and needs and other important features of the situation of use. Additionally, the testbed may include further social or technological conditions and circumstances, which are deemed crucial with respect to the usability or usefulness of the technological innovation. Like the technical prototype these other components, which together constitute the testbed, are prototypically realized.

"Prototypically realized" means that the actual implementation (be it physically or otherwise) of the technological and non-technological components is limited to those components and to those of their characteristics and features that are constitutive for the most relevant cause-effect relationships within the underlying scenario. Other characteristics and features are omitted, which the innovation as a marketable product necessarily would have to possess and that would be present as part of their real-world contexts of use. However, prototypically realized scenarios are associations of actually existing components with actual characteristics and with interactions between them that actually take place. The particular kind of environment provided by the laboratory makes it possible to realize situational scenarios without having to take into account all the requirements the envisioned future real-world application of the innovation would have to meet. The laboratory provides a containment against real-world conditions with which one otherwise would have to deal. Therefore, it allows constructing simplified versions of what would necessarily be much more complex as a real-world phenomenon.

The containment of the laboratory allows for simplifications that are obviously very helpful for getting a new technology up and running. However, engineers usually do not build technical prototypes just to show that these are working within a laboratory environment. Rather, the prototypical realization is thought of as only one step on the way to a new technology that eventually would count as an innovation in actual use.[3] Thus, a technical prototype that works within a laboratory environment is not an end in itself. It is valuable only because it indicates the possibility of a future real-world application and use of the technology represented by the prototype.

We believe that the capability of situational scenarios to serve as arenas of negotiation between the present and imagined futures is strongest in their manifestations as prototype scenarios. This is for two main reasons. The first reason has to do with the just mentioned relation between the prototypical realization of a technological innovation in the laboratory and the possibility of its real-world implementation. Because of this relation, every aspect of the prototype scenario in the laboratory refers to something outside the laboratory. Either it refers to actual states of affairs outside the laboratory, which are reproduced in a simplified manner within the laboratory. Or the prototype scenario refers to aspects of an imagined future reality outside the laboratory, which it anticipates in a nutshell

---

[3]This does not necessarily imply that the respective engineers would define it as their job to develop their prototypes this far. Most of the research engineers we studied see their work done with a prototypical realization of their new ideas. However, they frame their prototypical realizations as demonstrations of the performance future real-world implementations of their new technologies would have, thus confirming that a prototype that works within a laboratory setting is not meaningful in itself.

within the laboratory. In the first case, the relation between the reality within and the reality outside the laboratory is based on the assumption that the current reality will or shall stay the same with respect to these matters in a future where the new technology has become a real-world application. In the second case, it is based on the assumption that this future reality can be and should be changed according to the reality prototypically realized within the scenario. In the first case, the respective aspects of the prototype scenario are what we call present components, in the second case they are future components.

Because of the relation between the reality within and outside the laboratory, these present or future components are not just fictional elements of a fictional installation. Rather, they are connected to the real world: Either they represent aspects of the actual reality outside of the laboratory or they anticipate aspects of a future reality that is claimed to be attainable. This affects negotiation processes, because in order to support their positions, proponents of present as well as of future components can (and possibly have to) refer to the reality outside the laboratory. For instance, proponents of present components may point to deeply ingrained habits that will be difficult to change, while proponents of future components may emphasize the superiority of new or modified components by referring to deficiencies of current state of affairs. The main point here is that the prototype scenarios' reference to the reality outside the laboratory allows to and requires to substantiate the standpoints of the parties involved in the negotiation between present and future components. This possibility and necessity to substantiate the different positions is one of the main reasons why we believe prototypical scenarios to be especially useful and powerful arenas of negotiation.

The second main reason has to do with the prototype scenarios' character as materially realized situational scenarios. In contrast to the components of narrative scenarios, those of prototype scenarios are real entities with real properties and the interactions between them are actually taking place and not just in imagination. Thus, adapting a prototype scenario's components to each other means actually to manipulate or to influence their characteristics and their behavior. Figuring out how well the components are adapted to each other is a matter of actually testing the prototype within its testbed. Thus, the prototypical realization of situational scenarios provides a kind of reality check for the positions in negotiations between present and future components that a narrative scenario cannot provide. It allows recognizing and evaluating more clearly and more precisely the implications of one component's characteristics for other components' behavior and the associated costs of adaption. Admittedly, this reality check is provided by a simplified and fragmentary version of the reality the prototype scenario points at. Prototype scenarios do exclude many of the components and interrelations of the future real-world application, meaning that they do not allow dealing with all the problems of adaption related to them. However, to the degree that the protagonists take it seriously that prototype scenarios are meaningful only in relation to the future real-world applications they try to anticipate, this reality check can provide a quite effective means of structuring and coordinating negotiations between components.

5.1 Who Represents the Components in Negotiations?

Who does the negotiating when interaction and adaption between different components requires that one or another component needs to change their characteristics and behavior? Actor-network theory gives two different answers to this question. The first answer is that the elements of the emerging network themselves are negotiating their relations with others. According to actor-network theory, each of the entities that constitute an actor-network should be viewed as an actor, an actor being "any entity that more or less successfully defines and builds a world filled by other entities with histories, identities, and interrelations of their own"(Callon 1991, p. 140; cf. Latour 2005, p. 71). Thus, by trying to define or redefine other entities, the entities themselves negotiate their positions within the network. The second answer is that there are spokespersons who represent the standpoints and interests connected to those of the elements which themselves cannot voice them. Callon (1986, pp. 214–219) argues to this effect for instance in his case study on the domestication of scallops. He argues that as there are certain fishermen, which act as spokespersons of the fishing community, there are researchers acting as spokespersons of the scallops to be domesticated.

Both answers apply to the question of who represents the standpoints of present and future components in negotiations taking place in the arenas provided by prototype scenarios. In some respects, the components of a prototype scenario speak for themselves. For instance, the physical shape of a technical device that is believed to be functionally necessary or the taken-for-grantedness of a practice of everyday life may give the respective component of a scenario the power of resistance against change and thus makes them actors in the sense of actor-network theory. In this paper, however, we will focus on the second answer, that is, on the representation of the present and future components of scenarios by spokespersons. Our focus on human actors as spokespersons somehow is an implication of applying the arena approach to our topic, since arenas are sites of negotiation between social worlds and social worlds are groups of people with shared interests, practices and complexes of meaning.

What kind of members would be the ideal representatives of social worlds in negotiations concerning present and future components of prototype scenarios? First, they should be experienced members of the respective social world, assuming that a considerable part of the knowledge and practices of social worlds are tacit, which means that it takes some time to acquire them. Second, they should have a mandate to negotiate, that allows them (a) to speak for the social world they represent, (b) to mobilize the social world's resources to strengthen their position, and (c) to get the other members of the social world to accept and support the results of the negotiation.

In most of the research and development projects we studied, the representatives' mandates to negotiate are not given explicitly by the social groups they represent but is inherent to their positions and roles within the innovation projects. For instance, with the position as a research engineer working in an academic research facility comes a mandate that covers all three aspects just mentioned. Because of their position they are authorized (a) to represent the state of the art of the engineering discipline or specialty they are trained in, thus speaking for this field of research, (b) to use its knowledge resources and the devices and infrastructures of the research facility, and (c) to demand attention from the respective scientific community for their research results.

Much the same applies to test persons who play the part of the envisaged future users within a prototype scenario while themselves being members of this population. At least the first and third aspect of the mandate is inherent to their position as members of the social group that constitutes the imagined future users. This gives them (a) the authority to speak for them and their social world (or at least more authority to do so than other test users would have). And it provides them (c) with influence and credibility with other members of this social group when giving them their assessments of the risks and benefits of the new technology under development.

As mentioned above, the situational scenarios and their prototypical manifestations we studied in the field of ubiquitous computing are predominantly influenced by ideas and notions related to the social worlds of the engineering disciplines (or specialties) involved and of the envisaged contexts of application. Thus, we will focus on these two kinds of social worlds when we now consider how the perceptions and view of the social worlds are represented in the negotiations. The prototype scenarios, we studied, were built, specified, and evaluated by research engineers as part of their work in academic research laboratories. Because of this setting, the social world of the relevant engineering specialty (ubiquitous computing and related fields of computer science) is very well represented in the negotiations: by educated research engineers[4] with a mandate as described above that comes with their occupational and professional position. The nature and extent to which the relevant social worlds of the envisaged contexts of application are represented is much more diverse.

5.1.1   Representation by the Imagined Future Users

The just mentioned case of test users from the social group of the imagined future users is an especially strong and direct form of representation of application-related social worlds in prototype scenarios. This is true in particular, if the situation of use realized within the scenario is similar to an already existing real-world situation for which the new technology is developed. The paradigmatic example from our empirical data is prototype scenarios of in-house monitoring systems for seniors where real seniors volunteer as test persons. They live for some time in a smart home, which is furnished like a normal home but is equipped with a system of networked sensors for detecting movements. The task of the technological system is to detect unusual deviations from the inhabitants' daily activities, which would indicate that something is wrong with the person.

However, it is quite demanding to build this kind of prototype scenarios. On the one hand, it is costly and time-consuming to rebuild the corresponding real-world situation in the laboratory to a degree that it feels sufficiently real for the test users. And it is demanding to maintain these facilities—most of the smart homes we visited during our field research have become sights shown to academic visitors rather than still being testbeds that actually work and are used. On the other hand, it takes some effort to acquire test persons from the population of the envisaged future users. It is much easier to rebuild just those aspects of the intended context of application that are believed to be crucial and to employ as test persons students or researchers of the own research laboratory. Normally, the prototype scenarios we saw in our research were settings of this kind.

---

[4]Though it is common that PhD students do most of the work of constructing and coding the prototypes, there were usually also senior researchers involved in the technology development projects we studied.

5.1.2 Representation by Independent Expert Knowledge

Within such a setting, there are, again, substantial differences in nature and extent of how the application-related social worlds are involved. They are represented better than otherwise, when the situation of use in the prototype scenario is built based on valid knowledge about the corresponding real-world situation. In some of our cases, the research engineers tried to achieve this by referring to empirical knowledge about the relevant real-world situations. This approach, however, is limited by the fact that technology developers lack the necessary social science skills and that there are usually no social scientists or psychologists working at the research labs we studied.[5]

5.1.3 Representation by the Engineers Acting as Competent Users at the Same Time

Quite often, the research engineers of the projects and laboratories we studied solve the problem of being able to command over sufficient knowledge of the relevant contexts of application by choosing applications for which they themselves can count as competent users. Many situational scenarios are about ubiquitous computing applications for supporting office work, collaborative work with digital contents, networking between colleagues at conferences and so on. In addition, many scenarios depict solutions for everyday problems, thus referring to the everyday knowledge and competencies of the average person, for instance applications supporting users in not forgetting their door keys or applications that allow users to monitor their children playing in other rooms at home while working in their home offices.

5.1.4 Representation by Everyday Notions and Common Sense Assumptions

Yet another way to represent the practices, views and beliefs of the relevant context of application is to build the prototype scenario's situation of use based on

---

[5]Though several of the research engineers we interviewed, especially those who were heads of research laboratories, emphasized the necessity to include these competencies much more.

everyday notions and common sense assumptions. It is a common (and often tacit) practice in many spheres of life to fill in knowledge gaps by referring to common sense assumptions. The research engineers in our study proceed in this way when there is no direct or indirect knowledge about the context of application available to them or when it does not occur to them that they might need this kind of knowledge. Obviously, this is, if any, a much weaker form of representing the social worlds related to the context of use in the scenario than those mentioned before.

Nevertheless, even here the setting of a prototype scenario provides a kind of real-world test that helps to evaluate how reasonable these assumptions are. A prototype scenario is an interrelation between real entities with real properties. This applies to the aspects of the context of use considered in the scenario as well as to the technology. In a prototype scenario, common sense assumptions like any other knowledge about the context of use take the form of real entities with real properties if to become a part of the scenario. Thus, they have to prove that they are viable assumptions at least within the simplified reality of the prototype scenario. In this respect, a situational scenario that exists only as a narration or a theoretical construct and is not prototypically realized, always is a weaker form of representation. The prototype scenario's characteristic as an interactional setting between real entities with real properties provides a kind of reality check that those manifestations of situational scenarios cannot offer.

5.2 Who Represents the Present and Who the Future?

Based on the social worlds approach, we assume that a scenario's components and their interactions have different meanings for the representatives of the engineering specialties involved and for those of the context of application. The practices, understandings and beliefs constituting these social worlds guide their members to view and interpret certain components and their characteristics differently. Consequently, which components and features will win through and which will have to be adapted depends on how these different views are negotiated in the processes of building, specifying and evaluating scenarios.

Who, then, are the representatives of the present and of the future components? At first glance, it would seem that the representatives of the contexts of use would act as the defenders of the present components and the engineers would be the advocates of the future components. As a first approximation, this assumption is not all wrong, but needs to be qualified. It is most likely to be true (a) if the situational scenario depicts a future technological solution to a problem located within a current situation, which is assumed to stay more or less the same, and (b) if there is a rather direct representation of the relevant social world associated with this situation such as in the smart home scenario mentioned above. In such a case, we expect the test persons to defend their way of life and thus, at the same time, to defend the related present components of the scenario against change. Conversely, in order to push their new ideas forward the engineers will have to fight the resistance coming with this defense and act as advocates of future components. Additionally, we believe this to be the constellation in which the negotiation power of the defenders of present components is especially strong. Assumed that the future real-world context of application will be substantially similar to the related current situation, the approval or disapproval of the current situation's participants is probably the best hint available at what the envisaged future users might approve or disapprove. This is hard to ignore for engineers, if their long-term goal is to develop technology that gets used, even if it weakens their negotiation power in advocating future components. Conversely, this means that the position of user representatives would become weaker if the imagined future context of application differs from the related current situation. This raises the interesting question, who in this case may act as competent and strong representatives, an issue of some importance for prospective technology assessment.

Engineers aiming at technological innovations obviously are advocates of those future components that represent the crucial features of the new technology under development. Consequently, they will press for those modifications of a current network of components necessary for allowing these new components to find their place therein, and they will do so even if there are present components that have the strongest possible defenders on their side.

However, it is also true that technology developers act as defenders of present components. In the process of developing new technology, they often draw on already existing solutions from their own previous work or from the stock of their profession's state for the art. Many technology developers "seem to have a repertoire of them which they use over and over in their work" (Carlson and Gorman 1990, p. 393; cf. Jenkins 1984). In addition, nearly every contemporary technology development takes it for granted that a variety of technological infrastructures will somehow continue to exist. Conversely, it is perfectly possible that representatives of current contexts of application become advocates of future components. If they experience an aspect of their current situation as deficient and believe in the promise that a technological innovation will fix this problem, they may even be willing to sacrifice some of their beloved habits if necessary.

6 Empirical Findings

In this section, we discuss empirical examples for the different forms of representation of the social worlds related to the contexts of application in prototype scenarios. As we have argued, the social world of the relevant engineering specialty is usually very well represented by the engineers themselves. In contrast, the representation of the relevant social worlds of the envisaged contexts of application is much more diverse. Thus, our empirical focus will be on these different forms of representation. Above, we distinguished four different forms of how the practices, views and beliefs related to the relevant contexts of use are represented within the scenarios: (1) representation by the imagined future users, (2) representation by independent expert knowledge, (3) representation by the engineers acting as competent users at the same time, and (4) representation by everyday notions and common sense assumptions. It goes without saying that this is an analytical distinction. In our empirical cases, though one of these forms may dominate, we usually see combinations of them. Especially the last form of representation is present in virtually all of our cases. In the following sections, we discuss empirical examples for the first three forms of representation. Due to limited space, we will mention the last form of representation only in passing, as an aspect of the following empirical case.

6.1 Representation by the Imagined Future Users

We have already introduced prototype scenarios of in-house monitoring systems for seniors as the paradigmatic example of representation by imagined future users from our empirical sample. In the course of our empirical studies, we encountered several scenarios of this kind. This is not surprising, since monitoring systems for elders are a key application of the then emerging field of ambient assisted living (which is about using ubiquitous computing technologies to support elder people or people with special needs living normal lives). Our empirical example here is a prototype scenario built by ubiquitous computing researchers from a university on the east coast of the US. The underlying situational scenario assumes elderly people who live alone and who are still fairly independent. Their home is equipped with a system of networked sensors for detecting movements. The task of the technological system is to detect unusual deviations from the inhabitants' daily activities, which would indicate that something is wrong. In such a case, the system would notify a relative or a caregiver.

The research team's first step in moving into the field of ambient assisted living was to recreate a basic kitchen in the laboratory. They equipped it with video sensors monitoring the kitchen space from above, built a monitoring system and assigned to it the task of detecting cooking activities. Based on their everyday knowledge the engineers assumed that cooking is an activity that consists of a given sequence of basic actions that occur in a more or less fixed chronological order (getting the ingredients, preparing them, putting them on the stove, serving the dish). Consequently, the monitoring system would only have to detect the person's movements associated with these predefined sequences of action (from refrigerator to sink to stove to dining table) in order to detect cooking activities. In this prototype scenario, the interrelations of the present and future components reflect the rationale of the world of ubiquitous computing engineering while the context of application has little impact on the scenario. The engineers were focused on their technological idea to apply a "grammar-based formulation to detect complex activities" (Project publication 2006).[6] They built the prototype scenario in order to show that a "sensory grammar driven approach can invariably detect activity across many possible instances and actors" (ibd.). And they chose the task of detecting cooking activities simply because they assumed it to be a good example of an activity for which the grammar of its basic actions are easy to decipher.

Subsequently, the research team built a smart house, a normally equipped private home with a monitoring system that allows detecting the inhabitants' locations and movements. An elder person lived for more than seven months as a test person in this smart house. Later, the researchers continued their research with another smart home in which an elder couple with their adult son lived for several months (Project publication 2010). Obviously, the interrelations of the components depend much more on an adequate representation of the context of application in this scenario. The fact that the system's task now is to detect the activities of real people who are assumed to be similar to the future user, gives the context of application a much stronger negotiation position. This leads to major redefinitions of crucial technological components of the scenario.

Most important, the research engineers learned that their grammar-based approach does not fit with the actual behavior of the test persons, because most of the activities to be monitored do not show the predefined structure of activities

---

[6]We do not give the full reference of this and of all the following quotes from publications to maintain the anonymity of the researchers with whom we conducted interviews.

assumed by the grammar-based approach. Thus, they came to acknowledge that "very important information about the monitored person's habits [...] are often difficult to identify based on a priori or 'common-sense' knowledge" (Project publication 2009). The prototype scenario revealed that most of the regularities in the daily movements of elderly people at home are regularities based on individual habits. Thus, it became clear that the monitoring system at first would have to learn the inhabitants' individual habits in order to be able to detect unusual deviations from daily activities. Consequently, the engineers abandoned their original idea of monitoring activities based on a priori knowledge about sequences of action. They changed their technological approach and developed "an algorithm for determining if an event occurs persistently within an interval where the interval is periodic but the event is not" (Project publication 2010).

A large body of empirical research supports the social constructivist notion that there is no such thing as particular socio-technical constellations that inevitably prevail. Neither is there only one way of using a given technology, nor is there only one technological solution to a given problem. Why, then, does it seem so inevitable that the engineers had to abandon their original grammar-based approach and to adapt their monitoring system to the kind of behavioral patterns the test persons showed in the prototype setting? It is not the only solution to the misfit between the grammar-based detection strategy and the actual behavioral patterns of the users as represented within the scenario. Another solution would be to adapt the behavior of the users to the requirements of the monitoring system. However, the engineers accepted without discussion that this is not an option. Consciously or not, they assumed that seniors like those volunteering as test persons in their prototype scenario would not be willing to change their daily habits just to enable a monitoring system to detect them properly. In the negotiation arena provided by this prototype scenario, a present component (the seniors' patterns of daily behavior) won over a future component (the grammar-based detection approach). Consequently, the engineers replaced this future component by a different one able to adapt to this present component. This solution seems to be inevitable only because the negotiation position of the party associated with the present component is so strong that it seems pointless even to try to modify this component.

6.2 Representation by Independent Expert Knowledge

The case we chose to discuss the representation of the context of application by independent expert knowledge is also about technology development in the field of ambient assisted living. A research team from an Austrian university worked

continuously over several years on the development and evaluation of a technological system for automatically detecting falls of seniors in their homes under
real-world conditions. This monitoring system is intended to provide the basis for
a system that automatically alerts relatives or caregivers at a distance.

The prototype scenario the engineers developed and implemented in their laboratory consists of a set of different fall events and other movements (non-fall
events) conducted by members of the research team, the video cameras for creating 3-D-images of these fall events, algorithms for the analysis of the images
and thresholds for defining true or false fall events. The last aspect is of crucial
importance. To be useful in real-world applications, the fall detection system
not only has to detect true fall events reliably. With the same reliability, it must
avoid "false positives", that is wrongly identifying events as falls. Too many false
alarms would render the system useless in its intended context of use. The main
function of the prototype scenario was to assess (and improve) the accuracy of
different technological components (different sensors and different detection
algorithms) in detecting falls and distinguishing falls from non-fall events. A predefined set of fall events and non-fall events included in the scenario provide the
empirical test data for this purpose.

In the course of time, the engineers used this laboratory setting to develop and
assess a whole range of different technological solutions. After several steps of
evaluation of their initial technological system, which were not satisfying with
respect to the accuracy of fall detection (and especially the elimination of "false
positives"), the evaluation activities switched to different approaches. For example, to deal with the problem that the system frequently misinterpreted tables or
chairs as fallen people, the engineers added an overhead camera to better identify the shape of a fallen body. To provide another example: In one of the final
evaluation steps of the project, they used data from the overhead camera to add
a completely different kind of information to the overall detection system. The
engineers developed the idea to distinguish so-called "regions of interest", those
areas within the home, which the seniors are using more regularly than others.
They used the overhead cameras to detect movement patterns related to these
regions of interest for identifying non-normal situations, which then would serve
as an additional indicator for the occurrence of a real fall incident.

The social world of the context of application is primarily represented by the
set of fall events and non-fall events in the prototype scenario. For the following reasons we see this as an example for representation by independent expert
knowledge. The engineers obtain the set of movement patterns representing different forms of fall and non-fall events from a group of researchers from biomedical engineering who are specialized on this topic. In Noury et al. (2007) the

biomedical engineering experts develop a table of typical fall and non-fall events to serve as a concrete benchmark for fall detection (see Fig. 1). They argue: "At present it is practically impossible to compare the performances of different fall sensors from the data in the literature, as common criteria for their evaluation were not utilized nor were common procedures to carry out the tests adopted. We thus think that it is very important that objective criteria be adopted for the future evaluation of fall sensors with a framework for the evaluation of these devices"

TABLE I

SCENARIOS FOR THE EVALUATION OF FALL DETECTORS

Category	Name	Outcome
Backward fall (both legs straight or with knee flexion)	Ending sitting	Positive
	Ending lying	Positive
	Ending in lateral position	Positive
	With recovery	Negative
Forward fall	On the knees	Positive
	With forward arm protection	Positive
	Ending lying flat	Positive
	With rotation, ending in the lateral right position	Positive
	With rotation, ending in the lateral to the left position	Positive
	With recovery	Negative
Lateral fall to the right	Ending lying flat	Positive
	With recovery	Negative
Lateral fall to the left	Ending lying flat	Positive
	With recovery	Negative
Syncope	Vertical slipping against a wall finishing in sitting position	Negative
Neutral	To sit down on a chair then to stand up (consider the height of the chair)	Negative
	To lie down on the bed then to rise up	Negative
	Walk a few meters	Negative
	To bend down, catch something on the floor, then to rise up	Negative
	To cough or sneeze	Negative

Fig. 1 Table of fall and non-fall events (Noury et al. 2007, p. 1665)

(Noury et al. 2007, p. 1665). The engineers of our case study adopt this table as a benchmark for their fall detection technology. They use the events classified in this table as empirical test data for evaluating the performance of their different technological approaches.

These fall and non-fall events are in many respects abstractions from the real-world situations they represent. First, the empirical test data—fall (and non-fall) events that occur as specified in the table—are falls that do not really occur but are simulated by test persons. Second, the test persons are not members of the social world of the intended context of application. In this respect, the engineers of our case study follow the advice of the biomedical engineering experts: "Although the goal of a fall sensor is to detect the fall of elderly people, it is actually impracticable to test the fall situations with them. Thus the fall situations may be simulated by younger persons" (Noury et al. 2007, pp. 1665–1666). Third, the set of fall events specified by the biomedical engineering experts is not just a compilation of empirically occurring fall events but a typology based on theoretical reflections about how people fall. "As most falls occur during intentional movements initiated by the person, they happen mainly in the anteroposterior plane, forward or backward [...]. But in some cases, the fall occurs sideways, either during a badly controlled 'Sit-To-Stand' transfer, or if the person, when becoming unbalanced, tries to grip the wall" (Noury et al. 2007, p. 1665). Finally, as the descriptions of the fall events clearly show, the simulated falling behaviors of the test persons in the laboratory setting are strongly scripted. That is, the respective falling behavior impersonated by the test persons is prescribed to them in some detail.

The biomedical engineers' expert knowledge about fall events as it is used in the prototype scenario thus represents the social world of the context of application quite abstractly and indirectly. The target situation from the social world of application (falls of seniors in their homes) is translated in a laboratory situation that can be systematically processed in the social world of engineering research. Criteria for choice of people observed, definition and observability of falling behaviors and methodological feasibility are exclusively grounded in this social world's logic of sound scientific procedure. The fall (and non-fall) events serve as a benchmark for the fall detection system, providing the social world related to the context of application with a strong negotiation position—or a strong "voice". However, this voice can be uttered only very indirectly, within the laboratory setting into which the specificity of the application domain has been translated.

6.3   Representation by the Engineers Acting as Competent Users at the Same Time

Many of the situational scenarios and their prototypical realizations we learned about in our study envisage domains of application for which the engineers can count as competent users. In our analysis, this constitutes a form of representation of the context of application where the engineers themselves act as members of the group of the envisaged future users. Our example for this kind of representation is a location-based information system that was prototypically applied to serve as an intelligent audio guide for museums or art exhibits. The main interest of the Japanese research team which built this prototype scenario was to develop a technology that should embody a particular basic idea, namely to transfer useful information based on the position of the user in physical space without having to have an ID address to do so (Project publication 2002). The researchers did not intend to develop a technological solution suited for a particular domain of application. Rather, their goal was to develop a generic technology to be useful for a large range of different applications.

The technological principle that realizes the research team's basic idea is to transfer information via modulated infrared light to a hand-held (or ear-plugged) small device. On this device, solar cells receive the information encoded in the infrared light and a small processor transforms it into sound. A simple camera located at the spot where a particular information shall be provided (near an exhibition piece or a painting, for instance) detects whether such a device (and its user) is close by, and only then the infrared light transmits information. Because the infrared light in addition to the information also transfers the necessary amount of energy, the device needs no battery. Furthermore, the whole system is completely privacy preserving: "it is a non-ID communication, in the sense that it does not require ID for addressing. Physical location is used as the address" (Project publication 2002).

In the early stages of the development process, the engineers thought about many different possible domains of application such as situated information support at train stations or airports, providing information about traffic signals for blind people, or turning super markets into cyber marts. However, in all subsequent steps of development the engineers' focused on museums and art exhibitions as the context of application for their technical prototypes. Our empirical data from this case study do not really give an answer to why the engineers chose to focus on this domain of application. The engineers claim that their system "is ideal for low cost guidance terminal for museum exhibitions. It can provide location-dependent as well as personalized information, possibly tuned to the knowledge level and/or linguistic preference of the guest." (Project publication 2002). But very similar claims could be made with respect to a lot of other domains of application where such features would be welcomed as well.

The context of application they chose, however, has one definite advantage. Since the engineers themselves are visitors of museums and art exhibitions, they do not need to obtain expert knowledge in order to get to know the practices and orientations of their system's envisaged users. Nor do they have to win the support of some of the envisaged users to act as test users. Rather they can rely on their own expertise as visitors of museums and art exhibitions. This is what they actually did. From own experience they know that in a museum "a visitor usually wants to know the location of interesting exhibits or hear exhibit explanations" or that it "is difficult to provide custom-made information that fits each user, because interests often differ among visitors" (Project publication 2003).

The first prototypically realized laboratory setting that employs a prototype of the new technology imitates a museum booth (see Fig. 2). This version of the prototype scenario primarily serves the goals and criteria of the social world of engineering in that it mainly represents a proof of concept. However, though the imitated museum booth is just a very rough sketch of the museum situation it refers to, it nevertheless includes several components of current museum or art exhibitions and embodies suggestions about how to adapt the future information

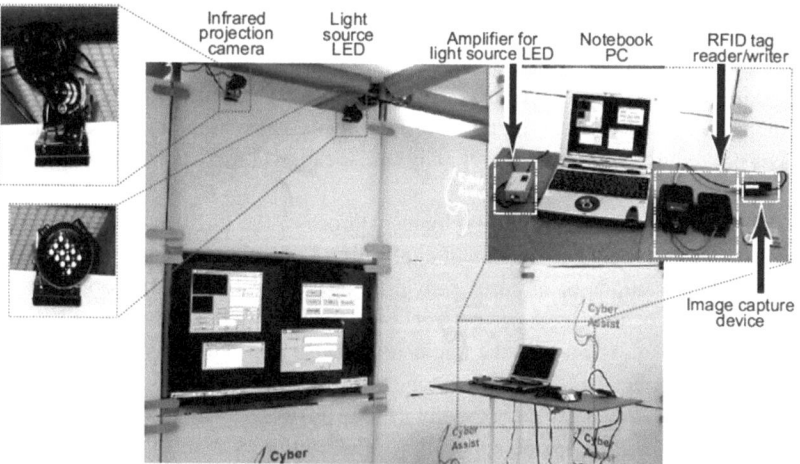

Fig. 2 Laboratory setting of a museum booth (Project publication 2003)

providing system to these present components. The present components to which the technological components are adapted in this prototype scenario include the positioning of wall-mounted exhibition pieces at eye level or the typical viewing distance from which visitors will look at the items at display. Obviously, it is sufficient that the engineers themselves are visitors of museums and art exhibitions to effectively represent these present components within the prototype scenario.

During the years the engineers worked on the location-based information system they built several other prototype scenarios. One of them took place at a real exhibition, the Expo 2005 in Japan. The main additional feature to prior prototypes of the system was that it should support providing information in different language. An international event like the Expo with its many international visitors is the obvious setting to make use of this feature. At the same time, it would exemplify the generic interactive potential of the system. The plan was to use different bandwidths of the same infrared emitter for a multi-lingual operation and to equip the listening device with a receiver able to switch between these bandwidths. However, with the technical components given and the time left until the opening of the Expo only a solution technologically less sophisticated was in reach. It was to install separate light sources for transmitting the information in Japanese and in English and to hand out two different listening devices, which were tuned into compute either the one or the other kind of light.

It is striking that the properties of the technical prototypes fit exceptionally well to the context of application provided by museums and art exhibitions. The location-based information system is virtually predestined for the task of providing visitors standing in front of exhibition pieces with audio information to be heard only by them. This is no coincidence, as it is the very reason the engineers chose this context of application. As mentioned at the beginning of this case description, the engineers' intention was not to develop a technological solution suited for a particular domain of application but rather a generic technology to be useful for a large range of different applications. For building a technical prototype and the related testbed they were free to choose from this range of possible applications. Thus, they chose a domain that satisfies two conditions: (1) to pose tasks that fit very well with what the new technology has to offer, and (2) to be well-known enough by the engineers so that they do not need external volunteers or expert knowledge in order to represent the context of application.

Many of our cases in which the engineers represent the context of application based on their own expertise share this characteristic. Prototype scenarios in which the engineers themselves represent the future users often are at the same time scenarios where the context of application is a setting that is chosen because it supposedly fits very well to the technological potential of the new technology.

This constellation has opposite consequences regarding the position of the context of application in negotiating the interrelations between present and future components.

On the one hand, it renders this position quite weak. Because if there turn out to be current features of the context of application calling for substantial change of the new technological components or even requiring an alternative technological approach, the engineers would rather change the domain of application than to do so. Since they chose the respective context of application because of its supposed quality to bring into effect the features that define the innovative potential of the new technology, it would make no sense to stick with it in such a case. Furthermore, changing the context of application is made easier by the fact that the engineers' investments in its prototypical realization are usually lower than when they had to acquire external expertise or volunteers from the group of the imagined future users.

On the other hand, the influence of present components from the context of application can be quite strong—as long as the features that set the new technology apart are not affected. As long as it serves to emphasize the innovative potential of the technical prototype, the engineers are eager to adapt their new technology to the requirements of current components of the context of application. This is why the engineers of the location-based information system welcomed the challenge posed by the international exhibition to provide information in different languages. They embraced this challenge as an opportunity to demonstrate the interactive potential of their technology.

7   Summary

Starting from the assumption that future images are vehicles for negotiation between the present and imagined futures, we developed the perspective of viewing situational scenarios as negotiation arenas. We argued, that in the case of situational scenarios the subject of negotiation is how to interrelate the components that make up the scenario. Situational scenarios are combinations of components from current situations ("present components") and of components that will come into existence only by realizing the imagined future ("future components"). The requirement to adapt present and future components to each other turns scenario-building (as well as all further specifications and evaluations of situational scenarios) into negotiating between the present and possible futures. This requirement is most explicit in the case of situational scenarios that are prototypically realized. Since the components of prototype scenarios are real entities with real properties and the interactions between them are actually taking place, adapting them to each other means actually to manipulate or to influence their characteristics and their behavior. Thus, prototype scenarios provides a kind of reality check for the positions in negotiations between present and future components that a scenario that exists only in imagination cannot provide.

In order to understand how these negotiations actually take place and what are the factors influencing the negotiation power of the parties involved, we referred to the arena approach of Anselm Strauss. Thus, we conceptualized the negotiation between the present and imagined futures as a negotiation of different perspectives related to the different social worlds, which are present in the negotiation arena. In our empirical cases, these are primarily the social world of the relevant engineering specialty and the social worlds of the envisaged contexts of application. We described the different forms of representation of the context of application we found in our empirical work on situational scenarios in ubiquitous computing projects. Based on empirical examples we discussed for these different forms of representation the kind and degree of influence of the social world of the context of application. While it is not a surprise that the negotiation position is stronger, the more direct the representation is, it is surprising to find that even rather indirect forms of representation are under some circumstances quite powerful in bringing to bear the perspective of the context of application. In our opinion, this is largely due to the characteristic of the prototype scenario as a structure that translates a future image into a provisional reality where even the most neglected of the components included have real effects. For this and the other reasons given or implied in this paper, we believe that prototype scenarios deserve much more attention from scholars interested in how socio-technical futures shape the present and thereby the future.

References

Borup, M., Brown, N., Konrad, K., & Lente, H. V. (2006). The sociology of expectations in science and technology. *Technology Analysis & Strategic Management, 18*(3–4), 285–298.

Callon, M. (1986). Some elements of a sociology of translation: Domestication of the scallops and the fishermen of St Brieuc Bay. In J. Law (Ed.), *Power, action and belief. A new sociology of knowledge?* (pp. 196–232). London: Routledge & Kegan Paul.

Callon, M. (1991). Techno-economic networks and irreversibility. In J. Law (Ed.), *A sociology of monsters: Essays on power, technology and domination, sociological review monograph 38* (pp. 132–161). London: Routledge.

Carlson, W. B., & Gorman, M. E. (1990). Understanding invention as a cognitive process: The case of Thomas Edison and Early Motion pictures, 1888–91. *Social Studies of Science, 20*(3), 387–430.

Clarke, A. E. (1991). Social worlds/arenas theory as organizational theory. In D. R. Maines (Ed.), *Social organization and social process. Essays in honor of Anselm Strauss* (pp. 119–158). Hawthorne: Aldine De Gruyter.

de Laat, B. (2004). Conditions for effectiveness of roadmapping. A cross-sectional analysis of 80 different roadmapping exercises. EU-US Seminar: New Technology Foresight, Forecasting Assessment Mathods, Seville, 13–14 May 2004. www.jrc.es/projects/fta/papers/Session%201%20Methodological%20Selection/Conditions%20for%20effectiveness%20.pdf.

Dierkes, M., Hoffmann, U., & Marz, L. (1992). *Leitbild und Technik. Zur Entstehung und Steuerung technischer Innovationen.* Berlin: Edition Sigma.

Geels, F. W., & Smit, W. A. (2000). Failed technology futures: Pitfalls and lessons from a historical survey. *Futures, 32*(9), 867–885.

Grin, J., & Grunwald, A. (Eds.). (2000). *Vision assessment: Shaping technology in 21st century society. Towards a repertoire for technology assessment.* Berlin: Springer.

Grunwald, A. (2009). Vision assessment supporting the governance of knowledge-The case of futuristic nanotechnology. In G. Bechmann, V. Gorokhov, & N. Stehr (Eds.), *The social integration of science. Institutional and epistemological aspects of the transformation of knowledge in modern society* (pp. 147–170). Berlin: Edition Sigma.

Grunwald, A. (2012). *Technikzukünfte als Medium von Zukunftsdebatten und Technikgestaltung.* Karlsruhe: KIT Scientific Publishing.

Grunwald, A. (2013). Techno-visionary sciences: Challenges to policy advice. *Science, Technology & Innovation Studies, 9*(2), 21–38.

Jasanoff, S., & Kim, S.-H. (2009). Containing the atom: Sociotechnical imaginaries and nuclear power in the United States and South Korea. *Minerva, 47*(2), 119–146.

Jenkins, R. V. (1984). Elements of style: Continuities in Edison's thinking. *Annals of the New York Academy of Sciences, 424,* 149–162.

Kahn, H., & Wiener, A. J. (1967). *The year 2000: A framework for speculation on the next thirty-three years.* New York: Macmillan.

Latour, B. (1988). Mixing humans and nonhumans together. The sociology of a door-closer. *Social Problems, 35*(3), 298–310.

Latour, B. (1991). Technology is society made durable. In J. Law (Ed.), *A sociology of monsters: Essays on power, technology and domination* (pp. 103–131). London: Routledge.

Latour, B. (2005). *Reassembling the social. An introduction to actor-network-theory.* Oxford: Oxford University Press.

Noury, N., Fleury, A., Rumeau, P., Bourke, A., Laighin, G., Rialle, V., & Lundy, J. (2007). Fall detection-Principles and methods. In IEEE (Ed.), *Engineering in medicine and biology society* (pp. 1663–1666). Lyon: EMBS 2007, 29th Annual International Conference of the IEEE.

Schaller, R. R. (1997). Moore's law. Past, present and future. *IEEE spectrum, 34*(6), 52–59.

Schulz-Schaeffer, I. (2013). Scenarios as patterns of orientation in technology development and technology assessment. Outline of a research program. *Science, Technology & Innovation Studies, 9*(1), 23–44.

Schulz-Schaeffer, I. (2014). Akteur-Netzwerk-Theorie. Zur Ko-Konstitution von Gesellschaft, Natur und Technik. In J. Weyer (Ed.), *Soziale Netzwerke. Konzepte und Methoden der sozialwissenschaftlichen Netzwerkforschung* (3rd ed., pp. 267–290). München: De Gruyter & Oldenbourg.

Schulz-Schaeffer, I. (2017). Akteur-Netzwerk-Theorie: Einführung. In S. Bauer, T. Heinemann, & T. Lemke (Eds.), *Science and technology studies. Klassische Positionen und aktuelle Perspektiven* (pp. 271–291). Frankfurt a. M.: Suhrkamp.

Schulz-Schaeffer, I., & Meister, M. (2015). How situational scenarios guide technology development-Some insights from research on ubiquitous computing. In D. M. Bowman, A. Dijkstra, C. Fautz, J. Guivant, K. Konrad, H. van Lente, & S. Woll (Eds.), *Practices of innovation and responsibility: Insights from methods, governance and action* (pp. 165–179). Berlin: Akademische Verlagsgesellschaft.

Schulz-Schaeffer, I., & Meister, M. (2017). Laboratory settings as built anticipations-Prototype scenarios as negotiation arenas between the present and imagined futures. *Journal of Responsible Innovation, 4*(2), 197–216. https://doi.org/10.1080/23299460.2017.1326 260.

Strauss, A. L. (1993). *Continual permutations of action*. New York: De Gruyter.

Strauss, A. L. (2016). *Creating sociological awareness. Collective images and symbolic representations*. New Brunswick: Transaction Publishers (First Publication 1991).

Sturken, M., Thomas, D., & Ball-Rokeach, S. (2004). *Technological visions. The hopes and fears that shape new technologies*. Philadelphia: Temple University Press.

Sydow, J., Windeler, A., Möllering, G., & Schubert, C. (2005). *Path-creating networks: The role of consortia in processes of path extension and creation*. In 21st EGOS colloquium, June 30-July 2, 2005. Berlin, Germany.

van Lente, H. (1993). *Promising technology. The dynamics of expectations in technological development*. Delft, Netherlands: Eburon.

van Lente, H. (2012). Navigating foresight in a Sea of expectations: Lessons from the sociology of expectations. *Technology Analysis & Strategic Management, 24*(8), 769–782.

van Lente, H., & Rip, A. (1998). The rise of membrane technology: From rhetorics to social reality. *Social Studies of Science, 28*, 221–254.

Ingo Schulz-Schaeffer (Ph.D. and habilitation) is Full Professor and head of the sociology of technology and innovation group at the Institute of Sociology, Technical University of Berlin. He is principal investigator of the DFG cluster of excellence "science of intelligence", and member of the program committee of the DFG priority program "digitalization of working worlds".

Martin Meister (Ph.D.) is a research associate at the Department of Sociology, Chair on Sociology of Technology and Innovation at Technical University of Berlin. Besides sociology and technology studies, actual interest is on technology assessment and social robotics.

Visionary Practices Shaping Power Constellations

Andreas Lösch, Reinhard Heil and Christoph Schneider

Abstract

The assessment of the impacts of sociotechnical futures on processes of innovation is a continuing challenge for technology assessment (TA). Imaginations of the future shape communications and actions in the present. They do not deliver any information about the outcome of these processes. This seems to contradict the task of TA to provide future-oriented knowledge. But vision assessment can provide knowledge concerning ongoing changes in sociotechnical arrangements and options for intervention. Our approach presents a heuristic for analyzing visions as socio-epistemic practices, which rearrange actors in innovation processes. In our cases—Big Data, Smart Grid and Fab-Labs—we identify dynamics in power constellations which might enable or restrain future innovations.

A. Lösch (✉) · R. Heil · C. Schneider
Institute for Technology Assessment and Systems Analysis, Karlsruhe Institute of Technology, Karlsruhe, Germany
e-mail: andreas.loesch@kit.edu

R. Heil
e-mail: reinhard.heil@kit.edu

C. Schneider
e-mail: christoph.schneider@mailbox.org

© Springer Fachmedien Wiesbaden GmbH, part of Springer Nature 2019
A. Lösch et al. (eds.), *Socio-Technical Futures Shaping the Present*,
Technikzukünfte, Wissenschaft und Gesellschaft / Futures of Technology,
Science and Society, https://doi.org/10.1007/978-3-658-27155-8_4

1   Introduction

The analysis and evaluation of impacts of sociotechnical futures, especially visions of the future on processes of innovation and transformation is a continuing theoretical, methodological and practical challenge in technology assessment (TA). This challenge is not easy to handle if one considers the task of TA to produce and provide future-oriented knowledge about the societal consequences of present processes of change. These processes are influenced by sociotechnical futures (such as visions, scenarios, roadmaps), which circulate in research policy, laboratory practices, the media and many more contexts. From a variety of research on visions (e.g. Grin and Grunwald 2000; Ferrari and Lösch 2017), imaginaries of the future (Adam and Groves 2007; Jasanoff and Kim 2015) and expectations (e.g. Brown et al. 2000) in TA and Science and Technology Studies (STS) we know, that such futures have various impacts on processes in the present. These processes could but do not have to lead to future innovations. The well-known problem of TA is that the futures, which circulate today in and between discourses and practices, are just imaginations of how the future could look like. They are "present futures" (e.g. Luhmann 1998), which are essential means for enabling communication and coordination of action between the heterogeneous actors involved in such processes (e.g. Lösch 2006). These sociotechnical futures do not tell us anything about "the future" (see also Grunwald, in this volume). TA cannot simply follow the claims of these visions to understand the consequences of research and innovation (e.g. for political decisions) but needs to create an understanding of the role of "present futures" based on present processes of change—that is processes, whose factual future outcome is unknown today. TA like any other science can only investigate how sociotechnical futures shape the present. Therefore the question remains open: How can TA be able to provide future oriented knowledge based on the shaping of the present by means of sociotechnical futures such as future visions?

In this chapter, we argue that for solving this problem we need to analyze the efficacy of future visions in ongoing processes of change. To enable this examination we need a specific analytical gaze for observing empirically what visions "do" in concrete practices. Therefore, we extend the focus of the "classical" vision assessment in TA from the analysis and evaluation of visions as ideas and narratives to an investigation of the practices in which visions are interwoven with other elements in *sociotechnical arrangements*. From this perspective, the meaning and the power of visions in such processes are not separable from their contexts; they emerge out of the transformations of the arrangements. Visions shape the present as socio-epistemic practices—that is, visionary practices produce sociotechnical arrangements and the knowledge needed for their changes simultaneously.[1] We consider these arrangements to entail discursive processes, practices, material and organizational aspects. With our approach we emphasize that the focus on visions can provide insights into the complex processes of multifaceted change within these arrangements. Depending on the different empirical scopes of our case studies we analyze such changes based on discursive dynamics and/or observable changes of the roles of actors, materials and organizational practices.

Our hypothesis is that the efficacy of visions as socio-epistemic practices in present transformations of arrangements is visible if we identify which responsibilities the visionary practices produce in the analyzed arrangements. If we follow distributions of *responsibilities* among the actors in correlation with the visionary practices we get insights on how they change their positions, roles and relations in the arrangement. Insights in these changes provide us with knowledge about the *power constellation,*[2] which enables the directions of potential innovations and transformations. With this application of our analytical concept "visions as socio-epistemic practices" on the investigation of processes of responsibilization in currently changing sociotechnical arrangements we contribute to the solution of the problem of TA—that is, to provide future-oriented knowledge although the future is only accessible in the dimension of the shaping of the present through "present futures".

In the following we unfold our analytical concept and position it towards related research in TA and STS. By means of three rather different cases we exemplarily show how visions shape and change present socio-technical arrangements practically.[3] In each case we focus on correlations between the visionary practices and their effects on the distribution of responsibilities among actors.

---

[1]The analytical concept of "visions as socio-epistemic practices" was elaborated during the first period of the ITAS-project "Visions as socio-epistemic practices. Theoretical foundation and practical application of vision assessment in technology assessment". See: https://www.itas.kit.edu/english/projects_loes14_luv.php. In the current second period we develop this concept further on.

[2]Our understanding of power constellations is oriented towards the relational concept of power of Foucault (e.g.; Foucault 1978, pp. 93–94). From this perspective options of influencing others, i.e. steering decisions and developments, depend upon the relational positions between all actors in an arrangement.

[3]A slightly different depiction of these cases, which does not reflect on the power constellations, is already published in Lösch et al. (2017).

The cases were chosen because they show that the work of visions as socio-epistemic practices and the correlating effects of responsibilization can vary significantly depending on the sociotechnical arrangement of the visionary practices. However, we approach the sociotechnical arrangements mainly from a perspective on the discourses about futures within them. Visions as socio-epistemic practices can and should be investigated in a more multifaceted manner and with a mix of empirical methods. In this chapter, however, we limit ourselves to exemplarily showing the changing of power constellations to highlight the potential of the heuristic approach that focuses on functions of visions—and as a starting point for further research within the respective fields and their particular ways of future-making. In the first case we identify dynamics of closing the future and stabilizing a traditional power constellation in expert and popular discourses on Big Data. In the second case we shed light on a rearrangement of power constellations in the electricity sector in experimental practices with smart grids in the context of the German Energy Transition. In the third case we identify the invention of new responsibilities in a short history of organizing FabLabs as arrangements of collective digital fabrication. We chose these three cases because the highly different power constellations that get visible through our analysis demonstrate the usefulness of our heuristic. Finally, we discuss the contributions of our specific vision assessment approach for solving the above-mentioned problem in TAs' knowledge production based on the insights from the cases.

2   The Analytical Concept in the Context of Related Research

Challenged by the systematic problem of TA's knowledge production Grunwald invented the vision assessment approach of TA in confrontation with the highly futuristic visions of nanotechnology (Grin and Grunwald 2000). Applications of this "classical" approach of vision assessment evaluated the contents of visionary ideas and narratives against the scientific and technological state of the art (the feasibility) and normative foundations of society (the desirability and acceptability). But to handle the systematic problem of TA this kind of vision assessment does not suffice. This problem is also reflected extensively by Grunwald in his previous publications, when he calls for a hermeneutic extension of TA, which should understand the assignments of meaning towards new technologies in communication processes in diverse arenas of society (e.g. public debates and controversies; Grunwald 2014). This extension of TA is an important task, because visions influence communications and actions, even if the contents and promises will never get realized (Lösch 2006). Such effects are assumed as being equal or more constitutive for processes of innovation and transformation than the factual realization of the visionary promises (implicit: Ferrari et al. 2012; Nordmann 2007).

In this dimension a variety of sociological studies on the role of visions and expectations has provided evidence that dynamics of expectations and anticipatory practices—e.g. the use of visions for guiding issues—correlate with dynamics of innovation processes (e.g. van Lente 1993; Konrad 2006; Alvial Palavicino 2016). Visions are seen as constitutive key factors which influence the processes and can at least enable innovations. These studies have also shown that the results of such long-term processes are explainable by reconstructions of stabilizations and destabilizations of heterogeneous actor networks in which visions play different roles, have different effects and are changing themselves (Borup et al. 2006; van Lente 2012). Although these STS perspectives are useful for the empirical demands of our research there are some important aspects and functions of visions in processes that have not been included in a way that satisfies the empirical needs of TA's vision assessment. In contrast to research which starts with the outcomes and reconstructs the initial dynamics in sociotechnical arrangements, TA and its vision assessment is committed to analyze and evaluate the role of visions in current sociotechnical arrangements in order to identify needs to intervene in such early stages of sociotechnical change. Although no one can know what the future will look like, for TA it is important to be able to identify the constellations in the present which create spaces of possibility (or impossibility) from which potentials of innovations and transformations could emerge.

Our analytical concept of visions as socio-epistemic practices is based upon the theoretical models and desiderata of this previous TA and STS research. Already in the 1990s, German social scientists coined the concept of "Leitbild" to describe visions associated with the design and implementation of new technologies. Through providing an integrative imaginary (the so-called "image function") "Leitbilder" serve as an activating, mobilizing and stabilizing reference point for the actors (Dierkes et al. 1996). From Leitbild-research we have learned that visions are capable to guide different actions in technological development. Later the sociology of expectations described how influential and powerful visions can be in debates about innovation processes (e.g. van Lente 1993; Brown et al. 2000). The strong orientation towards the future of visions and their broader and more ambiguous character (with respect to the examples analyzed by the German Leitbild theory), especially in the field of new and emerging technologies, resulted in the above mentioned "classical" vision assessment approach (Grin and Grunwald 2000).

Our analytical concept of visions as socio-epistemic practices develops these approaches further in order to enable TA to grasp the roles, functions and efficacy of visions as visionary practices within processes of change. Visions are not seen as representations of something. They have roles, functions and effects which emerge in practices and in interdependence to the sociotechnical arrangement, in which the practices take place. Different to many STS approaches our question is not if the visions get stabilized or destabilized; our leading question is, which arrangements do the visionary practices stabilize or destabilize.[4] We are asking what visions as practices produce. How can we analyze what the visions are doing? All the dimensions of such functions relate to each other or co-constitute each other in a functional manner, when visions serve as socio-epistemic practices. Through these new knowledge and sociotechnical arrangements are produced. We distinguish four important dimensions of practical functions of visions (similar on this topic see, e.g. Ferrari and Lösch 2017; Lösch et al. 2017; Lösch and Schneider 2016):

1. Visions *produce interfaces* between the present and an envisioned future. Such interfaces enable translations in discourses and actions between present constellations and the future and by doing so open up imaginative and practical possibilities (e.g. Adam and Groves 2007; Anderson 2010; Brown et al. 2000). In this dimension, visions provide orientation and enable the identification of options for change.
2. Visions *serve as communication media*, boundary objects or knowledge objects for different actors and discourses to which all the involved actors can refer, even if they have very different interests and perspectives (Lösch 2006; Star 2010; Knorr-Cetina 1997). In this dimension visions enable communication and interaction needed for change.
3. Visions can serve as *guiding visions* for actions, which practically enable coordination between different activities (e.g. Böhle and Bopp 2014; Dierkes et al. 1996). In this dimension visions produce coherence for the actors in the corresponding sociotechnical arrangements as an enabling condition of change.
4. Visions can activate behavior or changes of behavior of the addressed actors by unfolding a *normative force*. They position the envisioned and proposed innovations as desirable solutions for current and/or future problems or

---

[4]The vision itself is not understood as a stable and clearly defined entity. The vision is what the practices produce in reference to the imaginary point in the future. Similar for visionary images and discursive references as producers of their meanings see Lösch (2006).

challenges (e.g. Grunwald 2014; Nordmann 2010; Jasanoff and Kim 2015). In this dimension they address the specific responsibilities of the participating actors or even external actors in society who should do this or that following the visionary promises.

These practical functions of visions imply that we must make visible and explain how distributions of responsibility are produced during the (1) translations via the interface, (2) the communications enabled by the media, (3) the coordination of actions enabled by the guiding function, and (4) the normative activations of actors that visions unfold in practice. In this way, we can investigate which dynamic arrangements of distributed responsibilities are emerging through the work of visions as socio-epistemic practices and how these responsibilizations stabilize or destabilize the power constellation in the corresponding arrangement. The following three cases analyze such processes of responsibilization enabled and produced by visions as socio-epistemic practices.

3 Cases on How Visions as Socio-Epistemic Practices Shape the Present Differently

3.1 Big Data: Visionary Practices Stabilizing Power Constellations

Like every technological vision, the Big Data vision (Boyd and Crawford 2012; Schrape 2016) addresses not only technological changes, but also the necessity of social changes—that is transformations of the whole sociotechnical arrangement addressed by the vision. Therefore such changes modify power constellations between the actors and in turn imply a redistribution of responsibilities for the actors (Crawford et al. 2014; Morozov 2013).[5] In this case our insights on the changes of the arrangements related to Big Data are gained from the analysis of the use of the Big Data vision in discursive practices and its potential effects on strategies of action.

---

[5]The basis for this paragraph is a literature research in 2016 and the results of an expert workshop with participants from science, industry, politics, non-governmental organizations and data protection within the framework of the BMBF project "Assessing Big Data": https://www.itas.kit.edu/english/projects_grun15_abida.php.

In its temporal dimension Big Data is a vision of the future on the one hand, but on the other hand, it has already become a sociotechnical reality. As a socio-epistemic practice the vision performs as a fundamental means of orientation in discourses, thus constituting its own subject as it unites technological developments and challenges under a common term. In doing so, Big Data at least to a certain degree takes on the legacy of digitalization. The technical part of the Big Data vision is comparatively trivial: digitalization has progressed to a point where the resulting data is far too extensive, too complex and too heterogeneous for conventional methods of storage and analysis. The matter is in many cases further complicated by the requirement that the data should be processed in real time. This is the non-futuristic part of the Big Data vision rooted in the present, as it merely describes existing technical challenges resulting from current trends. But the Big Data vision also assigns those current trends to the future as a larger (social, economic, and political) horizon of meaning. On this level the vision produces an *interface* between the already running processes of digitalization and the future of Big Data. This interface orientates actions in the present towards an envisioned space of possibilities in the future.

As the vision classifies and interprets current trends in light of this horizon of meaning, the Big Data vision, seen as a socio-epistemic practice in the respective discourses develops a *normative force* of an activating imperative: According to the vision, data is not simply there and somehow needs to be processed; rather, its existence is considered to be something good and desirable. From this, an ethical obligation to produce more and more data is inferred for a variety of social actors. Through this normative postulate the vision develops an activation function that does not merely encourage certain patterns of action it almost "commands" them. Because the vision serves as a *communication medium*, which enables deliberations and negotiations between the participating actors, the normative imperative gets socially implemented—e.g., via distributions of responsibilities among the actors. Through this implementation of the normative imperative, the vision can serve as a *guiding vision*, which as a socio-epistemic practice develops specific principles to coordinate actions; the implementation of which it declares as a normative imperative of responsible action.

It is important to stress the point that the vision is not limited to improvements in data processing, but rather provides a legitimation for generating, collecting and processing additional data on one hand as well as the integration of existing data collections (Prodhan and Nienaber 2015) and their transfer to different or entirely new contexts of use on the other hand.[6] From the unchallenged and unconfirmed epistemological postulate, "More data and additional connections produce more knowledge," the vision infers a straightforward ethical obligation and moral responsibility: we therefore have to produce more data, improve the integration of existing data and use it in the most versatile way possible. Those who refuse to do so act unethically as they do not assume the responsibility that has been attributed to them. But how does this attribution of responsibilities work? Not only does the vision structure the field of argumentation in the discourse by providing concepts, it also defines what counts as legitimate action. It positions itself as essentially positive on one hand and as inevitable on the other. The vision does not deny negative consequences, but it presents them as mere side effects (that should be avoided as best as possible) of an essentially good and inevitable development. In short, the Big Data vision fuses (1) factual authority (those trends are existing in the present) with (2) a techno-determinist core assumption (those trends will continue whether we like it or not) and (3) a normative postulate (those trends are essentially good) into a powerful narrative.

Larry Page, one of the founders of Google, provides an example of this when he says: "Right now we don't data-mine health care data. If we did we'd probably save 100,000 lives next year" (Manjoo 2014). Page does not simply frame the evaluation of personal data by a private firm as a good thing; he implies that any behavior to the contrary is unethical, as it is inconsistent with the responsibility that the vision attributes to a potentially all-encompassing "we." Those who object to the vision are, by implication, potentially responsibility for the deaths of 100,000 people per year. Responsibility for Big Data is effectively delegated to society in the form of an imaginary "we." Therefore, only those who make their data available for the sake of a presumed enhancement of the common good and are willing to generate additional data act responsibly. Taken together, techno-determinism and diffuse attributions of responsibility for a postulated common good are thus functional complements of each other.

But attributions of responsibility as enabled by the Big Data vision lead to irresponsibility of the actors who actually participate in the collection, evaluation and commercialization of data. For instance, Big Data changes the perspective on the responsibility of businesses (Morozov 2013). The entrepreneur is presented or presents himself as making responsible use of his abilities and the revenue of his business in order to enhance society in accordance with the inevitable

---

[6]An example is the use of social network data for credit scoring (e.g. Wei et al. 2016).

requirements and possibilities of Big Data. In short, the entrepreneur as part of the collective "we" fulfills his collective responsibility as generated by the vision. It is getting evident, that the vision of Big Data works here as a socio-epistemic practice, which stabilizes the already running processes of digitalization and the power constellations between the actors, who profit economically from the collection, evaluation and commercialization of data and the collective "rest" of the society.

This effect of stabilizing and legitimating already established power constellations is likewise evident because the Big Data vision treats non-human entities in its sociotechnical arrangements, especially algorithms, as if they were responsible actors. This shifting of responsibility from human actors to artifacts is justified by the alleged objectivity of data and algorithms, on one hand, and the complexity of the digital world, on the other—a complexity that has become impossible for humans to control and that thus requires a delegation of decision-making to algorithms. The diffusion of responsibility within society as well as the attribution of responsibility to things (like algorithms) developed by the specific narrative of the Big Data vision as a socio-epistemic practice rules out traditional (individualizing) attributions of responsibility. Meanwhile, the fact that neither data nor algorithms are "objective" (natural facts) is neglected. The decision which data is or is not collected and in what way it is processed is already based on the preferences and interests of specific actors and thus inevitably flawed by values. The attribution of responsibility to a diffuse universality and the assertion of techno-determinist inevitability generated by the visions normative imperative exclude alternative social options of processing digital data, which are outside of the space of possibilities the Big Data vision creates.

In the addressed sociotechnical arrangements this vision stabilizes and legitimates existing power constellations: The groups of actors who are practically involved in the development and exploitation of technologies and processes accounted to Big Data by means of its vision are getting unburdened, if the collective of the society is made responsible for what they are doing. Therefore we can say that we see in the Big Data vision, the practical functions of visions interacting in a way that results in a continuation of present developments and power constellations, enabled by a specific mode of distributing and attributing responsibilities. In this case, we do not identify innovative changes in the corresponding arrangements. This mode of responsibilization differs significantly from the one the visions as socio-epistemic practices foster in the experimental changes with smart grid integrations in the course of the German Energy Transition.

3.2 Smart Grid: Visionary Practices Rearranging Power Constellations

Engineers and IT-experts describe the future smart grid as "a vision of a future electricity grid, radically different to those currently deployed, where the bidirectional flow of both electricity and information allows demand to be actively managed in real time, such that electricity can be generated at scale from intermittent renewable sources. [...] Unlike existing grids where electricity generally flows one-way from generators to consumers, [the smart grid] will result in flows of electricity that vary in magnitude and direction continuously" (Ramchurn et al. 2012, pp. 86–89).[7] Such a vision of a smart grid is very influential for the German Energy Transition and the transformations of energy systems worldwide because the vision serves as a solution for problems emerging in the course of the transition process and as a condition to enable desired changes (e.g. BMU & BMWi 2011, p. 19). In the case of the German Energy Transition, the vision of the smart grid is positioned as a future solution for the upcoming challenges to control and regulate an increasingly decentralized and dispersed energy system.

The vision addresses the increase in varieties of energy production and consumption, the volatility of renewable energies and the need for uninterrupted supply of energy. Different to the existing centrally controlled one-way system, which transports energy from particular sites of production to particular sites of consumption, the smart grid shall enable two-way flows of energy and information with sites of production and consumption changing in a decentralized system. A massive integration of digital technologies shall enable this. To test this vision, a variety of field experiments have been initiated and conducted in Germany and the EU with different experimental designs (Covrig et al. 2014; Nyborg

---

[7]The insights in this paragraph draw upon Lösch and Schneider (2016) where a more detailed analysis of an empirical study of smart grid visions in practice based on a document analysis and qualitative expert interviews can be found. The empirical work was conducted within the research project "Systemic risks in energy infrastructures", one of the projects of the Helmholtz alliance "Energy-Trans" (http://www.energy-trans.de/english/68.php). The document analysis included policy documents issued between 2007 and 2014 on the energy transition and smart grids mainly in Germany but also in the USA. Furthermore, it considered scientific texts on smart grid technologies from 1997 to 2014. In addition, a series of qualitative expert interviews was conducted in 2013. These included experts from power supply companies, an association of local utility companies, an industry association, an environmental association, a consumer protection association, technology companies and scientific experts, especially economists.

and Røpke 2013; Engels and Münch 2015; Goulden et al. 2014). Beyond the most general idea of making the electricity grid "smart" and increasing its automation, many different smart grid experiments have explored how this vision can be practically worked out. The overall smart grid vision links and integrates these different smart grid experiments.

One important and common insight gained through the German experiments is that the implementation of the principles depicted in the smart grid vision in the energy system would imply much more than new technical designs (e.g. BMWi 2014). The realization of a smart grid regulated energy system would imply far reaching and multidimensional transformations not only of technologies but also of all of the actors, the knowledge flows, modes of governance and much more. The central insight is thus that the whole sociotechnical arrangement of the electricity sector has to change. Envisioned changes include consumers turning to prosumers and inventions of new forms of electricity markets, novel business models and new sustainable everyday practices. Each of these changes would involve changes in everyday routines.

With such changes of positions and relations of the diverse actors in the transforming sociotechnical arrangements (tested in the experiments) a new power constellation and distribution of responsibilities among the actors would emerge. This is getting evident during the experiments, because here new distributions of responsibilities of the involved actors are tested and identified. Constitutive for such changes are the practical functions of visions as socio-epistemic practices to produce new knowledge and to enable the testing of new sociotechnical arrangements. This becomes clear if we look at expert evaluations of smart grid experiments. Experts from different sites of the energy system evaluate these experiments and do so in light of their interpretation of the vision of a future smart grid. Experts deploy the practical functions of the vision in order to develop their evaluations and their statements suggest that the visions are indispensable for the experimental practices and the resultant learning. Through the experts' use of the vision, the experiments created new responsibilities for actors in the electricity sector. The effects learned in the experiments, e.g. the identification of further experimental demands, were and are always accompanied by recognitions of newly demanded responsibilities and ascriptions of new responsibilities by the experts to actors, which in their view have not adopted the new responsibilities. Such ascriptions and adoptions of responsibilities are articulated by experts who helped to conduct or evaluate the experiments.

First, we can recognize that the visions produce an *interface* between the envisioned future smart grid and the present situation, with which the experiments were confronted. For example, an R&D manager of one of Germany's four 'big' energy suppliers points at new responsibilities of the energy suppliers to communicate with the customers. He mentions, that his "first surprise was how difficult it was to make customers participate. We had to explain endlessly what we want to do to inform customers why we wanted them in the field test. For us the system and the related changes were relatively clear, it's obviously not clear to the customer out there" (Energy Supply Company 2013). Second, we can say that for him and the other experimenters the vision served as an enabling *medium of communication* between themselves and the other actors in the experiments, mainly the customers. The vision as an interface and medium of communication furthermore produces insights in demands for changing the behavior of the customers (especially the business customers) in order to get "actively engaged" and to show a new kind of "flexibility" (Energy Supply Company 2013).

Third, his statement makes clear, that in the experimental situation the visions unfold their activating *normative force* by ascribing the responsibility towards the customers to learn to change their behavior and routines in orientation towards the vision of the smart grid. In turn, the old suppliers are getting responsible for enabling this learning process by initiating experiments. Here we see how the visions activate both actor groups to assume new roles. An imperative of behavioral change is emerging in the experimental practices, which does not only address the behavior of the customers, but at the same time the one of the suppliers. Fourth, we see here that the vision develops a coordinative power for actions in the sense of a *guiding vision*, which leads to the identification of an indispensable change of the modes of governance, which is getting plausible by the promised and experimentally proven controllability of a decentralized and volatile energy system under conditions of a future smart grid.

The experts' emphasis on an emerging and necessary new distribution of responsibilities in the sociotechnical arrangements of the electricity sector was enabled by the practical functions the smart grid vision unfolds together as socio-epistemic practices in the experiments. The new distribution of responsibilities is identifiable by observing the requirements for new sociotechnical arrangements that the experimental practices of the smart grid vision probe and produce.[8]

---

[8]As one consumer protection expert put it, "there is a completely different basic structure in the system and a regional responsibility and I think that is not yet properly communicated" (Consumer Protection Association 2013). Similarly, an environmental association expert states, "Now we're arriving at a certain stage where we want to turn the whole system upside down. There is an infinite number of actors that have to be included into the system" (Environmental Association 2013).

Statements by experts from different positions in the existing arrangements of the energy system show that visions, by modulating the experiments, enabled and produced processes of distributing responsibilities, which imply a transformation of the traditional power constellations. The visionary experiments point to specific modes of distributing responsibilities as a requirement for conducting further experiments and for translating the experimental results into the governance of innovation processes. The experiments produce the actors' responsibilities for future smart grids in present practices of their probing and learning from options for sociotechnical rearrangements.

We can identify a rearrangement of existing constellations, which correlates with the redistribution of responsibilities. According to the effects which the visions as socio-epistemic practices unfold in the experiments with smart grids, none of the actors in the energy system will be excluded or will lose their responsibility. But the responsibilities are changing. Some actors (suppliers, grid operators, regulators, etc.) remain responsible but have to extend and change their responsibility for new tasks and technologies. Other, more passive actors such as customers (both business customers and private households in their role as consumers) must take on new additional and active tasks (e.g. as prosumers). Therefore, we see a rearrangement of the power constellation in the electricity sector under the conditions of the smart grid vision. This process seems to both destabilize and stabilize the power constellation in the mode of more incremental transformations. The present is not extended by the future (as in the Big Data case), the present gets slightly reshaped (in the Smart Grid case). Both modes differ significantly from the inventory mode emerging in the organization of FabLabs.

3.3 Fab Labs: Visionary Practices Inventing Power Constellations

FabLabs ("fabrication laboratories") have become a global phenomenon of around 700 organizations that aim to provide simple and sometimes public access to digital fabrication, including 3D-printing.[9] They have been important in experiments with novel technologies in novel social contexts that are sometimes seen as

---

[9]This analysis is based upon an in-depth study of the emergence of FabLabs and the foundation of a grassroots FabLab that draws upon mixed qualitative methods such as participant observation, document analysis, action research, and interviews (Schneider 2017, forthcoming).

a "democratization" of technology and innovation (Walter-Herrmann and Büching 2013; Smith et al. 2016). In the FabLab case we show how the vision of Fab-Labs was transformed through socio-epistemic practices in material settings that linked FabLabs to other visions and in the processes changed the power constellations and the responsibilities of the actors in the respective arrangements. In particular, through such socio-epistemic practices of combining imaginations of the future and visionary practices, inventive responsibilizations have taken place and over time created actors that unfolded FabLabs.

The concept of FabLabs emerged at the Massachusetts Institute of Technology (MIT) at the Center for Bits and Atoms where researchers investigated 3D-printing and other computationally controlled processes to manipulate matter. This research was particularly influenced by visions of Nanotechnology that framed matter as formable based on single pieces, "atoms", which corresponded to the single pieces of digital information, "bits", that would provide the form. Furthermore, drawing parallels to the personal computer, the research was partly seen as contributing to the vision of "personal fabrication" that would allow "anyone to make anything, anywhere" (Gershenfeld 2012, p. 57) through machines that digitally transform matter. Through a funding program in 2001 that encouraged the "outreach" of science to society, the institute set up a handful of initial FabLabs with the aim to "deploy proto-personal fabricators in order to learn now about how they'll be used instead of waiting for all of the research to be completed" (Gershenfeld 2005, p. 11). These FabLabs were small workshops hosted by MIT or partner institutions that made relatively expensive, yet small-scale digital fabrication machines accessible to the public. They manifested a vision of an increasing capability of these machines to produce things and to become usable by anyone in the future.

In this first phase of FabLabs the vision created an *interface* between present and future that entails a particular configuration of responsibilities. The interface is established between existing digital technologies, personal computers and contemporary versions of digital fabrication machines that would both become more powerful in the future as their trajectories intersected. In the future, "anyone" is therefore seen as responsible for personal fabrication in which people make "anything," whereas the responsibility to reach this future is ascribed to research. This is, therefore, also enacted in the sociotechnical arrangements whereby MIT deployed FabLabs in public settings as prototypical arrangements of "personal fabrication." This was then still a rather typical power constellation with a corresponding distribution of responsibilities, which is based on a strong separation between research and society, between developers and users of technologies.

This rather constrained form of responsibility was to change a few years later. More FabLabs were set up with formal links to MIT and often funded through states and large organizations. In 2007, however, in the Netherlands an initiative of media and design organizations agreed with MIT to independently open their FabLabs without paying a fee or becoming a formal partner of MIT. And a few years later, also in the Netherlands, in 2010, the first "grassroots" FabLab was set up by a group of citizens, including artists and activists (Troxler 2014). By now hundreds of grassroots FabLabs, often run by volunteers, are providing access to shared digital fabrication machines. What has happened to enable this diversification of the organizational foundations of FabLabs, which is also a diversification of responsibilizations? In the decade that passed, two changes in the environment of FabLabs have taken place that entangled their history and vision.

First, in 2004, similar to the *guiding vision* of personal fabrication, the open source project RepRap started to develop open source 3D-printers and has since grown into an important hub for small-scale and inexpensive 3D-printing (Söderberg 2014). Initially, RepRap envisioned a radical transformation of the economy through its goal to make machines that make machines that would fundamentally decentralize the production of material goods. By now, most RepRap designs that circulate publicly are machines that produce small plastic objects. However, through the open source approach of the project, thousands of enthusiasts have positioned themselves as responsible for developing, using and experimenting with this form of 3D-printing and have thus opened a path of low-cost 3D-printing beyond industry (Tech et al. 2016). Typically, these open source 3D-printing enthusiasts, hobbyists and professionals alike also position themselves in relation to a vision of the emergence of a decentralized economy that they anticipate and experiment with in their practices (Alvial Palavicino 2016) and for which they consider themselves responsible for.

Second, and relatedly, the first decade of the 21st century has also seen more diffuse discourses on the "web 2.0" and related visions of "prosumers" and online collaborators that would lead to the "democratization of innovation" (von Hippel 2005). These discourses do not only draw upon older forms of Internet utopianism (e.g. Dickel and Schrape 2017a; Turner 2006) but also address "everyone" as a possible responsible agent in such processes. In relation to the production of material objects this diffuse discourse brought the figure of "the maker" into being that is seen as a prosumer who produces their objects empowered through Internet collaboration and typically uses 3D-printing (Dickel and Schrape 2017b). Here, we find suggestions that responsibilization made "makers" responsible for the further expansion of FabLabs.

The grassroots FabLab in 2010 was also based upon a normative vision of a new society through the Internet. In the following years, many more grassroots FabLabs emerged and conversely influenced the other visions, such as those just mentioned, as well. The diversity of FabLabs is also reflected culturally in many different goals that can be found within the FabLab landscape, entailing ideas of individual empowerment, novel forms of education or even a rejuvenation of industry (Schneider 2017). The vision that is part of practices to organize Fab-Labs thus unfolds a *normative force* as a socio-epistemic practice based upon desires to decentralize, democratize and to "open" innovation processes.

Within this environment, the visionary, yet manifest concept of FabLabs serves as a *communication medium* between different actors and practices. It offers a concrete example of how such novel organizations that make new and emerging technologies increasingly public and inclusive could look like. Precisely through such mediated communications, the vision of FabLabs addresses and integrates different actors and creates an arrangement based on a specific mode of responsibilization. FabLabs potentially address "everyone" as responsible in their further unfolding and FabLabs as organizations are responsible for reaching out to everyone. Since their beginning, the vision of FabLabs has enabled concrete FabLabs as experimental spaces to explore such visions and in turn transform themselves and the visions they were associated with (Schneider 2017; Troxler 2015). Through visionary discourse, organizational development and technical practices, this process of self-transformation also creatively unfolded novel responsibilities in relation to digital fabrication technologies. FabLabs thus appear to be a case of the invention of a new mode of distributing responsibilities that emerged through a creative process engendered through visions as socio-epistemic practices. The becoming distributions of responsibilities change the power constellation in the Fab Lab arrangements continuously.

At the present stage in the FabLabs we see the invention of a new mode of responsibility distribution, which is an innovation compared to the distribution of responsibilities in the arrangements of the first FabLabs at the MIT. But it is an open question how such internal changes interact with the external contexts of digital fabrication and other modes of organizing Fab Labs beyond the model of the grassroots FabLabs.

4   Conclusions and Consequences

The synopsis of the practical functions of visions shows in all of our cases how visions as socio-epistemic practices enable transformations of sociotechnical arrangements in present processes of change. Our focus on visions as socio-epistemic practices enriches the vision assessment of TA by providing an analytical view on visions in concrete and ongoing practices of rearranging the relevant sociotechnical arrangements. This corresponds with the challenges of TA to assess impacts, even though the future is unforeseeable and present constellations which show dynamics of change are the only starting point. The analytical concept is able to assist a process-accompanying real-time TA that tries to identify constellations from which potential innovations might emerge, including the critique of exclusions of alternatives in these processes, which might restrain the success of socially desirable innovations.

But in this chapter we did not only focus on the demonstration of the analytical value of our concept "visions as socio-epistemic practices" by means of our three cases. We went one step further and used the analytical concept as a heuristic to identify the changes in the distribution of responsibilities among the actors in the analyzed sociotechnical arrangements. Modifications in the distribution of responsibilities, which are created as results of the visionary practices, enable us to produce knowledge about changing power constellations in sociotechnical arrangements. To observe visions as socio-epistemic practices helps us to trace ongoing processes of responsibilization, which create constellations of responsibilities. We have shown how there are always dynamic processes engendered through visions that address particular actors as responsible or not responsible for particular positions or actions. These responsibilities stabilize or destabilize power constellations, which could be seen as the enabling or restraining forces of potential innovations.

The three exemplary cases showed how such responsibilizations can have different effects concerning the potential outcome of the changes. First, in the case of the Big Data vision we showed how this vision, through constructing a technodeterministic narrative linked to a presumed common good, renders responsibilities invisible and brings about "irresponsibilization" of the ones who are practically involved in the development and exploitation of technologies and processes accounted to Big Data, because the collective of the society is made responsible for what they are doing. The result is a closing of the future and a stabilization of the established power constellation and the current paths of digitalization. Second, in relation to smart grid visions and experiments, we observed and analyzed socio-epistemic practices to reform arrangements of distributing responsibilities with the aim to adapt the existing sociotechnical arrangement, including its dominant actors, to an emerging and still envisioned smart grid. Here we see an experimental rearrangement of the power constellation in the electricity sector under the conditions of the future smart grid. This process seems to simultaneously destabilize and stabilize the power constellation in the form of incremental innovations. Third, the analysis of FabLabs showed how a vision of personal fabrication over time combined with other visionary developments, increasingly turned the responsibility for the future of personal fabrication, away from science and research to decentralized constellations of responsibilities. In this sense, we see visions of FabLabs as socio-epistemic practices which invent new distributions of responsibilities together with the organizational invention of the corresponding sociotechnical arrangement. FabLabs thus appear to be a case of the invention of a new mode of distributing responsibilities that emerged through a creative process engendered through visions as socio-epistemic practices. In this case new power constellations and responsibilizations are invented by an organizational innovation.

Our analysis of these three cases provides fine-grained if limited insights into how visions as socio-epistemic practices produce distributions of responsibilities, which stabilize or destabilize power constellations in the respective arrangements. Due to practical, methodological and empirical limitations, we provide only situated rather than generalized views of responsibilizations and transformations of power constellations in these empirical cases. That said, the extensiveness of the constellations of responsibilities and power in innovation processes that we revealed here transcend individual research projects and furthermore, as we have shown, are constantly being worked on by heterogeneous actors and discourses. Therefore, desiderata of future research would be the analysis of the diverse correlations and dynamics between the changes in the observed arrangements (micro-level), processes of change in the contextual organisations and regimes and their power constellations (meso-level) and the intertwining of these dynamics with changes in the more general sociotechnical structures of our current society (macro-level) (see also the white paper in this volume).

References

Adam, B., & Groves, C. (2007). *Future matters: action, knowledge, ethics.* Leiden: Brill.

Alvial Palavicino, C. (2016). *Mindful anticipation: A practice approach to the study of expectations in emerging technologies.* Enschede: Twente University.

Anderson, B. (2010). Preemption, precaution, preparedness: Anticipatory action and future geographies. *Progress in Human Geography, 34*(6), 777–798.

BMU & BMWi. (2011). *The Federal Government's energy concept of 2010 and the transformation of the energy system of 2011.* Munich: Federal Ministry of Economics and Technology and Federal Ministry for the Environment, Nature Conservation and Nuclear Safety.

BMWi. (2014). *Smart energy made in Germany. Erkenntnisse zum Aufbau und zur Nutzung intelligenter Energiesysteme im Rahmen der Energiewende.* Berlin: Bundesministerium für Wirtschaft und Energie.

Böhle, K., & Bopp, K. (2014). What a vision: The artificial companion. A piece of vision assessment including an expert survey. *Science, Technology & Innovation Studies, 10*(1), 155–186.

Borup, M., Brown, N., Konrad, K., & van Lente, H. (2006). The sociology of expectations in science and technology. *Technology Analysis & Strategic Management, 18*(3–4), 285–298.

Boyd, D., & Crawford, K. (2012). Critical questions for big data-Provocations for a cultural, technological, and scholarly phenomenon. *Information Communication & Society, 15*(5), 662–679.

Brown, N., Rappert, B., & Webster, A. (Eds.). (2000). *Contested futures: A sociology of prospective techno-science.* Farnham: Ashgate.

Consumer Protection Association. (2013). Transcript of an expert interview with a spokesperson of a German Consumer Protection Association, conducted in 2013, quotations from the German language transcript translated to English by the authors.

Covrig, C. F., Ardelean, M., Vasiljevska, J., Mengolini, A., Fulli, G., & Amoiralis, E. (2014). *Smart grid projects outlook 2014. EUR-Scientific and technical research series.* Luxembourg: Office of the European Union.

Crawford, K., Miltner, K., & Gray, M. L. (2014). Critiquing big data: Politics, ethics, epistemology. *International Journal of Communication, 8,* 1663–1672.

Dickel, S., & Schrape, J.-F. (2017a). The renaissance of techno-utopianism as a challenge for responsible innovation. *Journal of Responsible Innovation, 4*(2), 289–294. https://doi.org/10.1080/23299460.2017.1310523.

Dickel, S., & Schrape, J.-F. (2017b). The logic of digital utopianism. *Nanoethics, 11,* 47–58.

Dierkes, M., Hoffman, U., & Marz, L. (1996). *Visions of technology. Social and institutional factors shaping the development of new technologies.* Frankfurt a. M.: Campus.

Energy Supply Company. (2013). Transcript of an expert interview with the head of an innovation group of one of the big German energy supply companies, conducted in 2013, quotations from the German language transcript translated to English by the authors.

Engels, F., & Münch, A. V. (2015). The micro smart grid as a materialised imaginary within the German energy transition. *Energy Research & Social Science, 9,* 35–42. https://doi.org/10.1016/j.erss.2015.08.024.

Environmental Association. (2013). Transcript of an expert interview with a leading expert for Renewable Energies from a German Environmental Association, conducted in 2013, quotations from the German language transcript translated to English by the authors.

Ferrari, A., & Lösch, A. (2017). How smart grid meets In Vitro meat: On visions as socio-epistemic practices. *Nanoethics, 11,* 75–91. https://doi.org/10.1007/s11569-017-0282-9.

Ferrari, A., Coenen, C., & Grunwald, A. (2012). Visions and ethics in current discourse on human enhancement. *Nanoethics, 6*(3), 215–229.

Foucault, M. (1978). *The History of Sexuality: An Introduction. Vol. 1.* New York: Vintage.

Gershenfeld, N. (2005). *Fab: The coming revolution on your desktop–From personal computers to personal fabrication.* New York: Basic Books.

Gershenfeld, N. (2012). How to make almost anything: The digital fabrication revolution. *Foreign Affairs, 91,* 43.

Goulden, M., Bedwell, B., Rennick-Egglestone, S., Rodden, T., & Spence, A. (2014). Smart grids, smart users? The role of the user in demand side management. *Energy Research & Social Science, 2,* 21–29. https://doi.org/10.1016/j.erss.2014.04.

Grin, J., & Grunwald, A. (Eds.). (2000). *Vision assessment: Shaping technology in 21st century society. Towards a repertoire for technology assessment.* Berlin: Springer.

Grunwald, A. (2014). The hermeneutic side of responsible research and innovation. *Journal of Responsible Innovation, 1*(3), 274–291.

Jasanoff, S., & Kim, S.-H. (2015). *Dreamscapes of modernity. Sociotechnical imaginaries and the fabrication of power.* Chicago: Chicago University Press.

Knorr-Cetina, K. (1997). Sociality with objects. Social relations in postsocial knowledge societies. *Theory, Culture and Society, 14*(4), 1–30.

Konrad, K. (2006). The social dynamics of expectations: The interaction of collective and actor-specific expectations on electronic commerce and interactive television. *Technology Analysis & Strategic Management, 18*(3–4), 429–444.

Lösch, A. (2006). Anticipating the futures of nanotechnology: Visionary images as means of communication. *Technology Analysis & Strategic Management, 18*(3–4), 393–409.

Lösch, A., & Schneider, C. (2016). Transforming power/knowledge apparatuses: the smart grid in the German energy transition. *Innovation: The European Journal of Social Science Research, 29*(3), 262–284. https://doi.org/10.1080/13511610.2016.1154783.

Lösch, A., Heil, R., & Schneider, C. (2017). Responsibilization through visions. *Journal of Responsible Innovation, 4*(82), 138–156. https://doi.org/10.1080/23299460.2017.1360717.

Luhmann, N. (1998). Describing the future. In N. Luhmann (Ed.), *Observations on modernity* (pp. 63–74). Stanford: University Press.

Manjoo, F. (26 June 2014). Larry Page on Google's Many Arms. *New York Times*, B1.

Morozov, E. (2013). *To save everything, click here: The folly of technological solutionism.* New York: PublicAffairs.

Nordmann, A. (2007). If and then: A critique of speculative nanoethics. *Nanoethics, 1,* 31–46.

Nordmann, A. (2010). Forensics of wishing: Technology assessment in the age of technoscience. *Poiesis & Praxis, 7*(1), 5–15.

Nyborg, S., & Røpke, I. (2013). Constructing users in the smart grid-Insights from the Danish eFlex Project. *Energy Efficiency, 6*(4), 655–670. https://doi.org/10.1007/s12053-013-9210-1.

Prodhan, G., & Nienaber, M. (2015). Merkel urges Germans to put aside fear of big data. *Reuters.* http://www.reuters.com/article/us-germany-technology-merkel-idUSKBN-0OP2EM20150609.

Ramchurn, S. D., Vytelingum, P., Rogers, A., & Jennings, N. R. (2012). Putting the 'Smarts' into the smart grid: A grand challenge for artificial intelligence. *Communications of the ACM, 55*(4), 86–97. https://doi.org/10.1145/2133806.2133825.

Schneider, C. (2017). *Transforming TechKnowledgies: the case of open digital fabrication.* PhD thesis. Munich, Germany, Technical University of Munich. https://mediatum.ub.tum.de/?id=1339289.

Schneider, C. (forthcoming). The becoming public of open digital fabrication. In S. Maasen, S. Dickel, & C. Schneider (Eds.), *TechnoScienceSociety. Technological reconfigurations of science and society*. Heidelberg: Springer.

Schrape, J. (2016). Big data: Informatisierung der Gesellschaft 4.0. *Berliner Debatte Initial, 27*(4), 12–21.

Smith, A., Fressoli, M., Abrol, D., Arond, E., & Ely, A. (2016). *Grassroots innovation movements*. Abingdon: Routledge.

Söderberg, J. (2014). Reproducing wealth without money, one 3D printer at a time: The cunning of instrumental reason. *Journal of Peer Production, 1*(4), 1–10.

Star, S. L. (2010). This is not a boundary object: Reflections on the origin of a concept. *Science Technology Human Values, 35*(5), 601–617.

Tech, R. P. G., Ferdinand, J.-P., & Dopfer, M. (2016). Open source hardware startups and their communities. In J.-P. Ferdinand, U. Petschow, & S. Dickel (Eds.), *The decentralized and networked future of value creation* (pp. 129–145). Dordrecht: Springer.

Troxler, P. (2014). Fab Labs forked: A grassroots insurgency inside the next industrial revolution. *Journal of Peer Production, 5*, 1–3.

Troxler, P. (2015). *Beyond consenting nerds: Lateral design patterns for new manufacturing*. Amsterdam: Hogeschool Rotterdam Uitgeverij.

Turner, F. (2006). *From counterculture to cyberculture: Stewart brand, the whole earth network, and the rise of digital utopianism*. Chicago: University of Chicago Press.

van Lente, H. (1993). *Promising technology. The dynamics of expectations in technological developments*. Dissertation, University of Twente, Enschede, Netherlands.

van Lente, H. (2012). Navigating foresight in a sea of expectations: lessons from the sociology of expectations. *Technology Analysis & Strategic Management, 24*(8), 769–782.

von Hippel, E. (2005). *Democratizing innovation*. Cambridge: MIT Press.

Walter-Herrmann, J., & Büching, C. (Eds.). (2013). *FabLab: Of machines, makers and inventors*. Bielefeld: transcript.

Wei, Y., Yildirim, P., van den Bulte, C., & Dellarocas, C. (2016). Credit scoring with social network data. *Marketing Science, 35*(2), 234–258.

Andreas Lösch (Ph.D. and habilitation) is sociologist, senior researcher and head of the research area "knowledge society and knowledge policy" and of the research group on "vision assessment" at the Institute for Technology Assessment and Systems Analysis at the Karlsruhe Institute of Technology.

Reinhard Heil is philosopher and researcher at the Institute for Technology Assessment and Systems Analysis at the Karlsruhe Institute of Technology.

Christoph Schneider (PhD.) is a sociologist and researcher with a focus on equitable and democratic digital futures, 3D printing and Responsible Research and Innovation at the Institute for Technology Assessment and Systems Analysis at the Karlsruhe Institute of Technology.

Part II
Socio-Technical Futures in Different Processes of Change

Globalizing Technologies: Geopolitical Innovation in the U.S. Bioeconomy

Tess Doezema

Abstract

This chapter examines how U.S.-based visions of biotechnological potential, undergirded by narratives of biological and economic progress, contribute to shaping technological innovation, policy, and politics at the international level. The invocation of "science-based" markets and policies is put to work in diverse instances to create conditions of potentiality for a particular vision of future good—the bioeconomy. The chapter discusses techniques for the creation of international markets for biotechnologies, as well as the modes of measurement and obfuscation leveraged in their validation and discursive construction as universally beneficial, placeless, and disembodied technologies.

1   Introduction

"The future is ours, but we have to planet"

"More Science, Less Politics!"
(Placards from March for Science in Phoenix, AZ 2017)

The March for Science in April 2017 drew hundreds of thousands of protesters across major U.S. cities and beyond (Smith-Spark and Hanna 2017). While the object of the march was to register dissent toward policies of the then recently

T. Doezema (✉)
School for the Future of Innovation in Society, Arizona State University, Tempe, USA
e-mail: tess.doezema@asu.edu

© Springer Fachmedien Wiesbaden GmbH, part of Springer Nature 2019   91
A. Lösch et al. (eds.), *Socio-Technical Futures Shaping the Present*,
Technikzukünfte, Wissenschaft und Gesellschaft / Futures of Technology,
Science and Society, https://doi.org/10.1007/978-3-658-27155-8_5

elected Trump administration that appears to base policy primarily on "facts" of the president's own creation, the characterizations of science displayed at the march demonstrated more than a public affirmation of the value of the variety of knowledge production practices broadly referred to as science. Slogans like those quoted above tap into a set of "widely shared understandings of forms of social life and social order attainable through, and supportive of, advances in science and technology" (Jasanoff and Kim 2015, p. 4). The declaration that "the future is ours, but we have to planet" conjures a perceived capacity of scientific expertise to calculate for and rationally order human affairs, and to harness the biological potential of the entire planet (whose existence appears threatened by the rejection of scientific reason) for the good of humankind. The second sign—"More Science, Less Politics"—rests on the premise that science and politics are diametrically opposed and fundamentally different phenomena; while positioning science as uniquely able to rationally order human society, allocate resources, and adjudicate the messy disputes currently mismanaged by politicians.

What these slogans suggest to be self-evident—how the world is rightly sorted into the boxes labeled "science" and "politics," and who the "we" is that must plan for the security of the planet—are key sites of struggle, where problems are framed and where the range of possible solutions to those problems is established in the process. Answers to these questions vary greatly across cultural and legal boundaries (Jasanoff 2005), and are worked out in popular scientific discourse, in court rulings, in regulatory rule making, and in innumerable acts of consumer choice and human interaction the world over. Invocations of a set of risks to the planet and its inhabitants that are understood to be global in scale, and thus warrant global solutions (Miller 2015), render the multiplicity of existing approaches to addressing these questions a source of conflict, as heterogeneity becomes an impediment to global progress. As national sociotechnical imaginaries collide on the global stage, the lines drawn to delineate science from politics become of central importance to how global cooperation is enacted. In this sorting process, particular forms of expertise are authorized to predict futures and prescribe action based on their privileged vantage points.

One broadly compelling narrative offered to address global risks is that of the bioeconomy. Bioeconomy plans and blueprints proliferate at the national and international levels—with national plans sprouting up from the U.S. to South Africa, and influential international organizations such as the Organization for Economic Co-operation and Development (OECD) and the European Commission affirming the promise and urgency of the nascent bioeconomy (European Commission 2012; OECD 2009). Despite the common use of the term bioeconomy, and the shared vision of a world redeemed by the power of bioinnovation, national bioeconomy plans differ importantly in how they define and seek to produce the bioeconomy (Staffas et al. 2013).

The U.S. bioeconomy blueprint depicts the bioeconomy not only as a national project, fundamental to "America's ability to innovate and compete in a global economy" (White House 2012, p. 35), but also a project of planetary redemption, key to supporting a global population of 8.3 billion people in 2030 (White House 2012, p. 10). This dual national-global benefit is emphasized throughout, in the form of affirmations that "leading the world in clean energy is critical to strengthening the American economy and addressing global climate change" (White House 2012, p. 10), and projections of the bioeconomy's capacity to "increase the potential for new and better medical products to promote a healthier American public and world population" (White House 2012, p. 30). *Existing technologies of the bioeconomy*—those technologies that must be sewn into the legal and regulatory frameworks around the world, and from which the future bioeconomy must grow—are framed as inherently beneficial by virtue of their role as precursors to future good, and their capacity to build the market infrastructure upon which the future bioeconomy depends. As I illustrate with the case studies of the AquAdvantage® (AA) Salmon and genetically modified (GM) crops below, markets for these technologies are laboriously produced through legal, discursive and diplomatic means. The array of cultural, legal, biological, and geographic negotiations that are involved in the production of existing technologies of the bioeconomy and their market-ability, are undertaken in the name of an imagined future dependent on biotechnological progress, as described by the Bioeconomy Blueprint. For the purpose of documenting the bioeconomy as viable, though, the struggles that produce and characterize these technologies are dismissed as messy, overly political distractions from the outsized benefits of the future bioeconomy that they portend. To put it differently, the cultural and material particularity of the production of technologies, and the work that goes into creating global markets for them, are systematically erased through modes of accounting and representation that measure volumes of production, trade and consumption. By this process, markets are made to appear as solely responsive to the merits of the placeless and disembodied biotechnologies in which they trade, and thus serve as clear evidence of presumed objective and universal technological merit.

In this chapter I suggest that despite the language of market liberalization in which the bioeconomy project is articulated, these dynamics represent more than yet another iteration of neoliberalism, or a new phase of capitalist development. This is rather a case in which two powerful narratives of progress, underwritten by two powerful kinds of expertise—biological and economic—are married and put in the service of one another. These are the mechanisms through which culturally, geographically and economically particular technologies are exported alongside institutional configurations and understandings of the world that are afforded a level of authority by virtue of their involvement in a broadly credible project of global technoscientific salvation.

Drawing on a diverse collection of empirical material including personal interviews, participant observation, publicly available documents from regulatory agencies, and diplomatic cables, this chapter examines the ways in which technologies of the bioeconomy are produced in geographically and culturally-bound ways, and simultaneously discursively constructed as universally safe and beneficial technologies for a global bioeconomy. This is a process of world-making that is underwritten by optimistic visions of bioeconomic potential. I offer case studies of the AquAdvantage Salmon and GM crops as exemplary technologies of the bioeconomy through which these dynamics can be observed.

2   Political Economy of a Bioeconomy

The bioeconomy is presented by its visioneers as the solution to globally imagined risks such as food, water and energy shortages, and conceptualized as a techno-economic system that is necessarily global in scale. The project is framed as requiring the participation of governments on all levels (OECD 2009). Indeed, while the term finds different meanings in different national contexts (Staffas et al. 2013), the proliferation of strategies reflects a remarkable level of consensus that the complementary realms of bioinnovation and market construction offer a dual pathway for addressing these global threats to human wellbeing.

A body of scholarship that analyzes the bioeconomy as a political or ideological project situates it as a continuation of historical processes of capital accumulation, either representing intensification of existing capital structures or amounting to a novel, perhaps more pernicious kind of capitalism (Birch 2006; Cooper 2008; Goven and Pavone 2015; Rajan 2006). Debating the novelty of adding the bio- prefix to economy, critical scholars of political economy characterize the bioeconomy as the continuation of ongoing processes of neoliberalization, suggesting that this powerful economic ideology is primary in shaping and driving the bioeconomy project (Birch 2006; Birch and Tyfield 2013; Cooper 2008; Goven and Pavone 2015). The reversal in causality offered by the argument that "a particular economic ideology—neoliberalism—has affected the bioeconomy rather than assuming that it is the inherent qualities of biotechnology that determine market value" (Birch 2006, p. 1), favors structural economic explanations over a certain kind of biological determinism identified in more traditional bioeconomic analysis (Carlson 2011; White House 2012), and offers as limited an understanding as that which it critiques. The bioeconomy project certainly draws on a range of economic expertise, and leverages the modes of measuring and accounting for the world that economics provides. Yet to label it a known (neoliberal) entity by virtue of its appearance as yet one more "marketization" scheme, reveals little about the world it seeks to create, the mechanisms of legitimation by which it proceeds, and the visions of good futures that animate them.

This analysis takes the bioeconomy project, instead, as a site where conflicting sociotechnical imaginaries of globalism are under active contestation and formation, allowing for insight into the ongoing creation and maintenance of an existing hegemonic sociotechnical imaginary of globalism. Globalism, in Miller's formulation, is a sociotechnical imaginary greatly shaped by scientists and international scientific institutions. It "transforms the Earth from a place that people live to a set of global systems that they inhabit and shape and that, in turn, imposes limits to which people must increasingly adapt themselves and their actions" (Miller 2015, p. 278). While the focus on global scale suggests that the imaginary of globalism is rightly also globally shared, the institutions within which this view of the world has been stabilized are not without their own power politics, values, and particularity of perspective. It must be noted the visionaries Miller sites as foundational to its rise are predominantly from the US and Europe. The sociotechnical imaginary of globalism Miller discusses must thus be understood as one (powerful) imaginary of globalism among possible others, which has become naturalized as *the* globalist imaginary through its coproduction with scientific modes of understanding and acting on the global systems in question. The U.S. bioeconomy blueprint proposes a global biological-economic system, clearly predicated on this hegemonic imaginary of globalism. Rather than existing in the world as a system already imposing its limits on global populations, though, it is posited as a possible *future* system, key to achieving future global security, the vision of which nonetheless suggests limits on the present global population, if we are to achieve it. Such future orientation renders the bioeconomy project an ideal site for understanding the continued intellectual and political work upon which a seemingly unified imaginary of globalism depends, in its aspiration to the kind of global cohesion and unity that its proposed techniques of measurement and intervention presume. I argue here that the U.S. bioeconomy project is a dynamic site at which boundary work delimiting politics from science is actively undertaken by a wide range of actors in order to stabilize and sustain a particular imaginary of globalism, which remains actively contested within and beyond U.S. borders. Boundary work is carried out in an array of spaces in

which the authority to claim resources, make decisions, and take action relies on
the delineation of the scientific from the political, mechanical, religious, social or
otherwise (Gieryn 1983). As such, boundary work plays a key role in maintaining
social order and shoring up institutional practices that rely on an idealized separation between science and values (Jasanoff 1990). In the context of the bioeconomy project, particular demarcations of science from values, politics, and culture
is essential to establishing shared understandings of the nature and purpose of a
global bioeconomic system, and the expertise necessary to act on it in an objective, scientifically valid manner. At stake are the ways in which these boundaries
get drawn, whose logics hold power within the resulting borders, and with what
consequences for democratic modes of governance and self-determination.

Understanding sociotechnical imaginaries of globalism as particular and contested, despite their universal lens on the world, makes legible the U.S. bioeconomy project's dual agenda aimed at maintaining American competitiveness,
and at global transformation of trade, markets, energy production, manufacturing,
medicine and food production. This paired national/global focus implies on one
hand that what is good for America is good for the world, but also infuses the
blueprint with a sense of urgency, in that to maintain competitiveness, we must
outpace our global competitors who would write the rules in less virtuous ways.
The urgency of the bioeconomy exists within the context of confounding public
dissent to particular biotechnologies to which no small amount of theorizing, ire
and social engineering aspirations have been directed (e.g. Juma 2016; Specter
2009). This tension between a sense of urgency for bioeconomic progress and the
threat of political dissent undermining the achievement of global transformation
is illustrated by a key instance from President Obama's 2015 State of the Union
Address:

> As we speak, China wants to write the rules for the world's fastest-growing region.
> That would put our workers and businesses at a disadvantage. Why would we let that
> happen? We should write those rules. We should level the playing field. That's why
> I'm asking both parties to give me trade promotion authority to protect American
> workers, with strong new trade deals from Asia to Europe that aren't just free, but
> fair. Look, I'm the first one to admit that past trade deals haven't always lived up
> to the hype, and that's why we've gone after countries that break the rules at our
> expense. But ninety-five percent of the world's customers live outside our borders,
> and we can't close ourselves off from those opportunities… 21st century businesses
> will rely on American science, technology, research and development. I want the
> country that eliminated polio and mapped the human genome to lead a new era of
> medicine-one that delivers the right treatment at the right time. In some patients
> with cystic fibrosis, this approach has reversed a disease once thought unstoppable.

Tonight, I'm launching a new Precision Medicine Initiative to bring us closer to curing diseases like cancer and diabetes—and to give all of us access to the personalized information we need to keep ourselves and our families healthier. (Obama 2015)

Here barriers to innovation are posited as arising from international disharmony that threatens to impede the ability of U.S. bio-innovation to realize its full potential. President Obama seamlessly integrates global competition, trade, and bioinnovation as natural parts of an argument that identifies existing forms of democratic ambivalence surrounding global trade as inhibitory to the kind of progress necessary to bringing about a brighter biological future. He weaves a narrative in which the promise for improved medicine through biotechnological achievement is at risk in the absence of widely contested trade agreements that he argues would level the global playing field for American businesses. Appealing to shared hopes for improved health and wellbeing and a thriving economy, he presents these as naturally dependent on particular global economic configurations. Obama appeals to Congress and the nation to allow him to write the correct kinds of rules in international agreements—not for the sake of free markets or deregulation, but in the name of bio-innovation and unlocking the potential for global biological transformation on the scale of a "moon shot" (Obama 2016).

The bioeconomy project draws on a discourse of pressing contemporary crises in terms that produce a world that is amenable to the application of particular kinds of expertise. Expert knowledge and prescriptions for ordering the world do not go unchallenged, though. The registers in which opposition to expert agendas in biological and economic world-making has been expressed, and the points of disagreement among experts and key publics become learning moments for both sides, shaping future strategies and ways of conducting research (Epstein 1996). Likewise, technological design is in many ways, an act of shaping political and social relations (Hecht 1994), enabling biotechnological innovation to anticipate and address dissenting views as part of design principles and choice of technologies. Similarly, regulatory regimes can also be designed to anticipate and perform responsiveness to perceived public concern about technological innovation. A closer examination of market-making dynamics surrounding existing technologies of the bioeconomy elucidates ways in which biological and economic narratives of progress are leveraged in tandem to create the conditions of possibility and justification for a bioeconomy project that is made to seem beyond reproach and universally beneficial. Here I take the optimistic pairing of the biological with the economic seen in bioeconomy plans and blueprints as an invitation to examine the techniques of suturing these domains of expertise together, as illustrated by two examples of existing technologies of the bioeconomy.

3   Pioneering Technologies

The AquAdvantage (AA) Salmon is and has been widely characterized as an exemplary technology of the bioeconomy. It is a fish that is genetically engineered to grow to market size twice as fast as its non-engineered brethren, and represented as capable of contributing to solutions to ever increasing global food demand. The fish underwent a twenty-year regulatory journey through the United States Food and Drug Administration (FDA), emerging in 2015 as the first genetically engineered animal to be approved for human consumption—only to be immediately entangled in a lawsuit brought by environmental groups (Institute for Fisheries Resources 2016). Congress has further delayed the passage of fish to plate by way of a rider to a spending bill pausing the sale of AA Salmon in the U.S., pending production of labeling guidelines by the FDA.

The imagined potential of biotechnology—material, economic and social—that underwrites the U.S. Bioeconomy Blueprint (BEB) vision is evident in the regulatory process of the fish itself, if in far more subtle form. The FDA prefaces its Environmental Assessment (EA) of the AA Salmon, noting the bioeconomic necessity and potential of the technology under assessment by highlighting problems of food scarcity, unsustainable fishing practices, and increasing global demand. "World-wide demand for protein production has increased significantly in the past decades… and fish protein often comprises a significant portion of the daily dietary protein in many countries" (United States Food and Drug Administration 2012, p. 6). It also notes that cold water fin fish are a source of low fat protein and Omega-3 fatty acids, juxtaposing U.S. health and nutrition needs with increasing global demand and corollary depletion of wild fish stocks (United States Food and Drug Administration 2012, p. 6). Thus, FDA too characterizes the salmon as a potential solution to problems of U.S. public health, global food supply, and unsustainable exploitation of natural resources, noting a necessity to "meet increasing demand for fish protein in light of declining stocks and diminishing capture of wild fish" (United States Food and Drug Administration 2012, p. 6). The FDA's account renders as fact this framing of a global health crisis derived from the simple laws of supply and demand and necessitating the increased production of fish protein. This depiction of the world, articulated in the language of economic science, becomes part of the factual background environment on which the fish's "impacts" are assessed in regulatory review. As such, they are incorporated into the constructions of risks and benefit that FDA is tasked with evaluating and the body of "sound science" (United States Food and Drug Administration 2015) upon which it renders its regulatory decisions. While FDA vigorously maintains that it "makes science-based decisions, and doesn't consider the economic implications of our decisions to approve or not approve products" (Personal communication, June 21, 2016), the agency's sense of the stakes of regulation, and of the social and economic import of its judgments for the future of the bioeconomy are integral to how "sound-science" is defined.

The significance of establishing the right kind of regulatory precedent in the case of the AquAdvantage Salmon is echoed by an open letter addressed to president Obama and signed by a group of self-described "concerned international scientists and global technology company executives." Preceding the FDA's decision to approve the Salmon for human consumption, they argued, "the obvious regulatory roadblocks AquaBounty is experiencing not only undermine our ability to meet the future food needs of the world, but seriously damage the global credibility of FDA and its objective, science-based approval process, while stifling innovation in this critical field" (Letter to President Obama 2014). Action is required in the present, they argue, in order to create a regulatory precedent that will not only allow this particular technology to enter U.S. markets and the American diet, but to foster innovation for a host of "as yet unimagined" (White House 2012, p. 2) technologies that will feed the burgeoning world population. The use of the term "science-based" performs a rather delicate form of boundary work, as it prescribes the removal of the decision from what the concerned scientists and technologists call "misplaced economic/marketplace concerns and reactionary fears of people who either don't understand or choose not to understand the science behind the AquAdvantage Salmon" (Letter to President Obama 2014), while in the same gesture loading the regulatory decision with geopolitical and humanitarian significance, deemed by expert judgement to be directly relevant to the science of regulation. These visionaries demand the performance of "more science, less politics!" on the part of the Obama administration, in the process explicitly delineating what rightly falls into these two categories.

As in the BEB, discourse surrounding the engineered salmon figures U.S. interests and global benefit as aligned and overlapping. The U.S. is posited as an ideal leader in biotechnology innovation and regulation, poised to seize the opportunity to write the rules for biotechnology regulation through a recognizably "science-based" process, such that they are universal and can thus be exported to the rest of the world. The FDA is framed as a de facto institution of global governance, capable of scientifically determining correct forms of regulation for biotechnologies—or of botching the job—such that global needs can be met by a harmonized system of risk-based regulation. Bioeconomy advocates repeatedly posit this role for U.S. regulatory frameworks, even while criticizing the lengthiness of regulatory review processes. Calling the fish "pioneering," one commentator opines, "if the U.S. government moves to review and update a rigorous and efficient approval process for future animal biotechnology products, it would define model regulatory standards and best practices for the rest of the world" (Juma 2016). This comment positions FDA in a key role for market-making, suggesting that the agency's task is not solely to keep dangerous substances out of the U.S. food supply, but also to open, engender and set the parameters for markets. To put it differently, the FDA is presumed to act on the definition of science-based reason offered above, taking into account particular aspects of the geopolitical and humanitarian significance of its decisions, which have been deemed part of a science-based assessment of biotechnology, while narrowing its purview in other key ways that allow for the silencing of forms of dissent not based on this vision of a good future.

The work of rendering biotechnologies placeless and disembodied is predicated on this synecdochical role of the individual technology—the AA Salmon—as representative of the promise of the future bioeconomy. Biotechnologies must be disentangled from the particular bodies in which they reside, the environments that sustain them, and the local customs and laws that produce and regulate them. That is to say, in order to be able to cross legal and ecological boundaries around the world, they must be extracted as cleanly as possible from their biological entanglements with the world. This depends on a process in which "good governance" is engineered directly into biotechnologies, such as through built-in bio-containment measures of the AA Salmon (Hurlbut 2018). AA Salmon are an all-female and (99.8%) sterile population, rendering them incapable of multiplying outside of the lab, or breeding with wild populations should physical containment measures fail. They are raised in fully contained land-based facilities, further extracting them from the complexity of the ecosystems and troublesome migration patterns in which their conventional counterparts are engaged. In this way, desirable features of the biological are extracted and standardized, while the unruly and unmanageable aspects of these living bodies are designed out of the technologies such that they can be transferred freely through the world as commodities, extracted from particularities of body, land and law.

The neat and orderly story of science-based regulation for placeless and disembodied technologies is belied by actual practices of regulation. The FDA negotiates regulatory jurisdiction of the AA Salmon by displacing responsibility to other jurisdictions. AquaBounty R&D is based in the U.S., with egg production and fertilization in Prince Edward Island, Canada, and cultivation ponds in Panama. The tripartite production chain allowed the FDA a dual expansion and contraction of responsibility. On one hand, the demands of containment and monitoring requirements of the biological status of the fish gained the FDA access to oversee industrial production and containment processes beyond the U.S.

boarders: "The FDA will maintain regulatory oversight over the production and facilities, and will conduct inspections to confirm that adequate physical containment measures remain in place" (United States Food and Drug Administration 2015). On the other hand, the international production chain precluded the need to assess the impacts of the potential failure of that containment as part of the EA, since the failure of containment in Panama would presumably not have effects on the U.S. environment. The EA states to this effect that its approval only permits "production and grow-out of AquAdvantage Salmon at facilities outside of the United States," and thus, since "the areas of the local surrounding environments that are most likely to be affected by the action lie largely within the sovereign authority of other countries," the agency is not bound to consider environmental effects of the AA Salmon, "except insofar as it was necessary to do so in order to determine whether there would be significant effects on the environment of the United States" (United States Food and Drug Administration 2012). With this delicate negotiation of responsibility, the agency constructed a highly situated and geographically specific regulatory regime around the fish.

For all the promise of universal application of sound science, the AA Salmon only satisfied the requirements of safety by trading on localized authority and limited jurisdiction. "The environment," so easily figured as a global construct in biotechnological promises, became someone else's responsibility by virtue of siting production elsewhere. AquaBounty's response to the approval of the fish in the U.S. and Canada, though, reaffirms the presumed universality of these assessments of safety as untethered from political or geographic particularities. Ron Stotish, AquaBounty CEO states, "Health Canada authorities gave their approval for our salmon to be produced, sold and consumed in Canada, following a similar approval from the U.S. Food and Drug Administration in 2015. We now have independent scientific reviews and approvals from two of the most respected regulatory agencies in the world that *have confirmed the safety of our salmon for human consumption and for the environment*" (Towers 2016, emphasis added). This statement returns the AA Salmon to its status as a truly placeless technology, declared safe by the world's leading science-based regulatory agency, and ready to reach grow out tanks, markets, regulatory approvals and dinner plates across the globe.

## 4	Cultivating Global Markets

The open letter to president Obama, cited above, explicitly likens the AquAdvantage Salmon to genetically engineered crops as evidence of the merit of the fish and the role of biotechnology in fostering U.S. bioeconomic prowess: "The technology used to develop the AquAdvantage Salmon is the same technology used to develop the biotech crops planted on over 350 million acres around the globe" (Letter to President Obama 2014). They go on to highlight the technological leadership played by the U.S. in both developing and commercializing the technology, claiming that "Biotech crops have enabled American farmers to become the most productive farmers on the planet; this technology continues to contribute to making agriculture one of the most dynamic and important sectors of our economy" (Letter to President Obama 2014). Here the sheer volume of biotech crops around the world, as well as their significant contribution to the U.S. economy is the primary and seemingly obvious evidence of their merit. The BEB similarly points to genetically engineered crops as central to the success and promise of the bioeconomy, looking forward to advances "expected in the near future to lead to crops with other desirable traits such as improved nutritional value, enhanced disease resistance, and higher crop yields," and promising to "move this mainstay of the bioeconomy to a new level" (White House 2012, p. 11). While their present value is demonstrated by market dominance, their future potential is articulated as tangible, practical, and unambiguous human benefit. Genetically engineered corn, soy and cotton are positioned as phase one technologies of the bioeconomy, lending legitimacy to the prospect of the AquAdvantage Salmon through a common technological heritage, and demonstrating the global potential of the bioeconomy. The technology used to develop GM crops and the AA Salmon is represented as having been discovered in the U.S., but capable of delivering undifferentiated global benefit, offering enhanced capabilities to farmers around the world and capable of "making balanced meals available to all" (Monsanto). Rob Carlson, a biotechnology researcher/lecturer and consultant whose bioeconomic analyses are cited in the blueprint itself to demonstrate the economic power of bioinnovation,[1] enthusiastically describes the rapidly expanding market share of the bioeconomy: "revenues within the U.S. are presently about US$ 125 billion, or approximately 1% of GDP, and growing at a rate of 15–20% annually. While this torrid pace will ultimately slow, it is clear neither when this will happen nor how large a fraction of U.S. GDP biotech could ultimately provide" (Carlson 2007, p. 110). This form of analysis keys into the bioeconomic promise of new wealth created through nothing more than human ingenuity and the power of "rational engineering of biology" (Carlson 2007, p. 116).

---

[1]Carlson was also referred to as the best source of empirical data on the bioeconomy by one of the authors of the U.S. National Bioeconomy Blueprint, during a breakout session of the workshop on Research Agendas in the Societal Aspects of Synthetic Biology, November 2014, Tempe, AZ.

While burgeoning markets provide the basis to evidence the benefits of bio-technologies, significant labor on the part of the U.S. government goes into the production of those markets around the world. A leaked State Department memo from 2007 laying out the Department's "FY 2008 biotech outreach strategy" defines a set of goals. One of these is "to facilitate trade in agbiotech (sic) prod-ucts by promoting understanding of the technology and encouraging the adoption of fair, transparent, and science-based policies and practices in other countries. Another important goal is to promote understanding of biotechnology as a tool for supporting economic growth and improving food safety and security in devel-oping countries" (United States Secretary of State 2007). The Department defines these goals as explicitly global, while identifying a set of predominantly devel-oping countries (Brazil, Burkina Faso, China, Colombia, Czech Republic, Egypt, Germany, Ghana, India, Indonesia, Kenya, Nicaragua, Nigeria, Peru, Philippines, Romania, Russia, South Africa, Thailand, Ukraine, Vatican, Vietnam) as targets of particular focus (United States Secretary of State 2007).

Cables back to the department from its various posts around the world illus-trate the host of strategies undertaken by the Department's representatives to upend mandatory labeling schemes, to partner with local NGOs hosting pro-bio-technology conferences, to educate foreign leaders about the benefits of biotech, and other strategies aimed at influencing lawmaking, the ways in which regula-tions are enacted, consumer preferences, choice of crops planted on farmers' land, etc. Throughout these documents references to "science-based" policies, approaches, facts and principles arise repeatedly, operating as a central concept of the geographically diverse efforts toward bioeconomic construction. To pre-vent mandatory labeling of GM foods in Hong Kong, a diplomatic post proposes outreach programs targeted at legislator's slated to review the labeling guidelines "designed to provide stakeholders with facts of biotechnology using a science-based approach" (Consulate Hong Kong 2008). The implication is that scientific expertise can offer a clear understanding of what information consumers have the right to know about their food, and that these principles are universal—invariable from Kansas to Hong Kong.

A cable from Cairo notes that "Monsanto also has been collaborating with the Agricultural Research Center of the MOA for five years to develop Biotech cot-ton seeds. These cannot be sold in Egypt until the government puts into place a workable, science-based regulatory system" (Embassy Cairo 2006). This state-ment suggests that bioscientific knowledge provides a clear basis upon which markets should be constructed, and that regulatory regimes are rationally deriv-able through scientific means, and accordingly will foster the unhindered progress of science and technological diffusion.

Another cable reports on the visit of a senior U.S. agricultural biotech advisor to Turkey, "to discuss the current status of Turkey's biosafety regulations and to promote science-based, pro-development policies in this area" (Embassy Ankara 2005). The diplomat reports, "her outreach to the press, business, and parliamentarians, was very useful in helping Post broaden the vigorous debate on the draft law and educate potential stakeholders about how their interests could be advantaged through biotechnology or disadvantaged through a restrictive, non-science based regulatory process" (Embassy Ankara 2005). Here science is represented as a natural ally of development, while regulations that are too limiting and that do not advance biotechnology in the country are framed as unscientific. In the strategic use of science and biotechnology as the basis of regulatory reform throughout the world, science takes on a meaning that is at once grand, universally benevolent and unassailable, while also carrying geopolitical significance and mandating very specific alterations in how regulators are asked to think, what laws are appropriate to progress, and what information consumers around the world should have access to.

Production and success of novel organisms is framed in the bioeconomy blueprint as an imperative worthy of public support and protection in order to overcome "market failures" (White House 2012, p. 2) that might result in less than optimal levels of production and implementation, and their success around the world is granted generous diplomatic and financial scaffolding by the U.S. State Department. Yet their ultimate uptake on farms around the world and consumption by increasing numbers of consumers is ultimately attributed to their inherent technological superiority. In the face of mixed evidence of the beneficial nature of genetically modified crops, Carlson declares "Given the variability in assessing crop performance, I feel the best indicator of the farm scale benefits of GM crops is simply the continued use and increased adoption by farmers worldwide" (Carlson 2011, p. 2). This method of assessing benefits is the State Department's favored method of assessing the utility of the GM crops around the world as well. An offhand comment from a diplomat posted in Slovakia surmises, "The acreage increase among fewer farmers suggests the science and effectiveness of GMOs has been compelling for farmers who have stayed with the product" (Embassy Bratislava 2009). This use of market expansion to demonstrate the benefits of GM crops produces a circular logic. The U.S. government supplements the creation and commodification of novel life forms because of their presumed superiority to their non-engineered counterparts, their predicted social benefits and their inherent vulnerability in the face of market forces that produce biotechnological progress at a rate deemed less than "optimal" (White House 2012, p. 2). At the same time, success is finally measured not by how they perform in producing predicted social benefits, but rather by their dominance in the market. That market is pains-takingly constructed on the premise of universal "science-based" policies and expert predictions of future benefits. Increased trade in biotechnology is then rep-resented as the result of market forces neutrally responsive to the merits of the technologies it trades in, and thus as a source of further evidence of the universal benefits of biotechnology.

This account elides the role of U.S. law, regulation and diplomacy in creat-ing the conditions of possibility for the market success of GM seed in the U.S. and around the world. Particular regulatory regimes support the spread of GM crops, while others—deemed "unfair" and inadequately attentive to "the science-based principles and consumer benefits of biotechnology" (Consulate Hong Kong 2008)—favor the success of other cultivars. The introduction and market satura-tion of GM corn and soy in the U.S. was dependent on court rulings that reflect cultural understandings of where the boundaries between nature and invention lie (Jasanoff 2012). It relied on a regulatory regime in which no new regulatory laws were thought to be necessary in order to govern the release of novel forms of life (Jasanoff 2005, p. 52), supported by agricultural subsidies that incentivize the production of staple crops, and by an American diet heavily reliant on a processed food industry that takes advantage of subsidized corn and soy to create cheap, standardized food products. As GM crops are promoted around the world to cre-ate global markets for a bioeconomy, corresponding efforts are necessary not only to change regulations and intellectual property law in a multitude of national con-texts, but to contend with other more subtle consequences of the particularities of biotechnologies that were created in U.S. fields: for large scale farms and farm-ing machinery, for farmers with the tacit knowledge to apply the required inputs, the cultural aptitude to adopt such technologies and negotiate contracts with seed production companies, and for an American population accustomed to consuming industrially processed food products. This work is elided in measurements and reports of a burgeoning bioeconomy trading in placeless and disembodied tech-nologies, proliferating by virtue of the power of "rational engineering of biology" (Carlson 2007, p. 16).

The narratives that frame biotechnologies as disembodied and placeless, and narratives of markets as universal mechanisms of ordering and validation are engaged in tandem for these projects of science-based market-making. The the State Department's communications display a lexicon and a mode of reasoning in which market-making practices are simultaneously understood to be "sci-ence-based" since correct laws and regulations are understood to be empirically derived—as placeless as the technologies they regulate—and insofar as they are conducive to creating a system that will be supportive of the multiplication and diffusion of biotechnologies "as yet unimagined" (White House 2012, p. 2). To put it differently, markets are referred to as "science-based" in two, complementary ways—they are (1) understood to be derivable from universal natural law, and (2) able to unleash a flow of scientific and technological progress. The vision of an abundant future brought about by science-based markets and market-based science tightly links practices of U.S. national regulatory decision-making and U.S. science-diplomacy around the world. Though optimistic and geared toward increasing innovation, this vision of a good global future has a resultant limiting effect on present democratic possibilities for decision-making around the ways in which particular technologies do and do not foster environments for human flourishing, and how regimes of ownership and modes of exchange in biological goods might serve present human needs and diverse aspirations for the future.

5   Making and Measuring Markets

While propagating existing technologies of the bioeconomy relies on particularities and inconsistencies of local law, culture and jurisdiction to patch together markets and governance regimes, such projects trade on notions of the universality of scientific reason as both the justification and the validation of their efforts. Existing technologies of the bioeconomy—such as GM crops and the AA Salmon—are treated as disembodied devices of market-making, where the messy particularities of their physical bodies are irrelevant, since their value lies not in the specificities of their interactions with the world, but in their ability to lay the way for the full promise of the future bioeconomy. This renders any possible objections a citizen might have in the present to these technologies irrelevant and inhibitory to progress, no matter how real and grounded in experience they may be. To be scientifically minded about biotechnologies means being keyed to a broad vision of future promise, not to overly political embodied controversies of the present.

Despite the soundness of regulatory science affirmed at every juncture of the movement of the AquAdvantage Salmon from lab to plate, the conditions for its approval rely on delicate boundary work to delineate politics from science, and spatially and geopolitically dispersed allocations of authority and responsibility. The conditionality of the approval of the fish for the U.S. marketplace is elided in discourse around science-based regulatory decision-making and the capacity of this technology and others like it to address global hunger. Similarly, for all the public affirmations of the scientific consensus around the benefits of GM crops, their increasing profitability and proliferation relies heavily on the

U.S. government as an agent of market creation. This takes the form of creating "science-based" policies abroad, presumed to invite the kinds of innovation deemed necessary to redeem the world from looming global risks. Measurement of the environmental, health and social effects of existing biotechnologies, such as Roundup Ready corn and soy, are essentially a non-issue for bioeconomy visioneers, as the numerous, tangible and disparate effects of existing technologies are secondary to their function as tools of geopolitical innovation and global market making such that the fruits of the future bioeconomy and biotechnologies as yet unimagined can be globally ushered in through markets that have been prepared for them in the present.

Carlson's predictions, calculations, and arguments against regulating access to tools of biotechnological advance are—like those of the bioeconomy blueprint, the regulators at the FDA, and the State Department officials who tirelessly spread the good word about biotechnology around the globe—based on this same promise of future biotechnological benefit that the bioeconomy project insists on. But his project of conjuring the bioeconomy into a measurable reality while ostensibly providing objective data on its current status, relies on a continued blindness to the "deep and often invisible politics of what is counted, why it is counted and how the counts are used" (Hilgartner 2007, p. 385). The project of counting revenue and acres of GM crops around the world, which elides the ways in which that expansion has been produced, serves to shore up what its calculations omit, and reinforces the narrative of the naturally-expanding bioeconomy, running on the dual engine of human ingenuity and as-yet-untapped biological potential.

In the pursuit of regulatory reform for the bioeconomy throughout the world, U.S.-based actors, like their ideological counterparts at the Science March, rhetorically position science as grand and universally benevolent. The broad stroke of declaring a particular set of sociopolitical relations "science-based" draws and hardens political battle lines across a divide that delineates as scientific a set of aspirations for a future world bound up in the imagined possibility of economic and technological progress. Such delineation impoverishes public deliberation around the possibilities, compromises and payoffs of biological technoscientific production, and possible modes of material exchange and distribution. With the polarization of political discourse along the lines of scientific believers and unbelievers, more nuanced debate over what technologies and what public purposes are worth pursuing appears less and less possible. Accordingly, the appeal to scientific objectivity to legitimize broad projects of political and economic change undermines the very power that one hopes to draw in declaring an agenda aligned with scientific reason, and in the process, alters the shared significance of what we understand the word *science* itself to mean.

The bioeconomy project as the projection of a particular sociotechnical imaginary of globalism rests centrally on the representation of biotechnologies as universally benevolent, unbound from particularities of local environments, biological bodies and local politics, and produced by the unerringly rational forward march of scientific progress. Yet the world into which these technologies enter—and which makes them possible—is one where power and authority are uneven and negotiated and where responsibilities are easily displaced. It is, in Ulrich Beck's words, "a world of organized irresponsibility" (2009)—a world whose workings suggest that the effects of efforts to create a global bioeconomy will be plural, not singular, chaotic, not ordered, haphazard, not rational, and whose benefits will be partial and ambivalent, not unequivocal. The enduring conviction that the biological productivity of the planet can be harnessed by scientific reason, and saved from destruction through processes of planning, ordering and genetic improvement, continues to shape technological innovation and conceptualization of ways forward for global governance in critical and largely unacknowledged ways.

Acknowledgements Many thanks to Ben Hurlbut for guidance and careful reading of early drafts of this chapter, and to the editors of this collection for their insightful comments.

References

Beck, U. (2009). *World at risk*. Cambridge: Polity.

Birch, K. (2006). The neoliberal underpinnings of the bioeconomy: The ideological discourses and practices of economic competitiveness. *Genomics, Society and Policy, 2*(3), 1–15.

Birch, K., & Tyfield, D. (2013). Theorizing the bioeconomy: Biovalue, biocapital, bioeconomics or what? *Science. Technology & Human Values, 38*(3), 299–327.

Carlson, R. (2007). Laying the foundations for a bio-economy. *Systems and Synthetic Biology, 1*(3), 109–117.

Carlson, R. (2011). Biodesic 2011 bioeconomy update. https://obamawhitehouse.archives. gov/sites/default/files/microsites/ostp/biocon-%28%23%20001SUPP%29%20Biodesic_2011.pdf. Accessed 11 Aug 2011.

Consulate Hong Kong. (2008). Funding request for FY2008 biotechnology outreach and capacity building in Hong Kong. *WikiLeaks Public Library of US Diplomacy*. http:// wikileaks.wikimee.org/cable/2008/01/08HONGKONG186.html. Accessed 29 Jan 2008.

Cooper, M. (2008). *Life as surplus: Biotechnology and capitalism in the neoliberal era*. Seattle: University of Washington Press.

Embassy Ankara. (2005). Senior biotech advisor's meetings with Turkish officials and agribusiness. *WikiLeaks Public Library of US Diplomacy*. https://wikileaks.org/plusd/ cables/05ANKARA862_a.html. Accessed 15 Feb 2005.

Embassy Bratislava. (2009). BT corn acreage increases despite administrative and commercial obstacles. *WikiLeaks Public Library of US Diplomacy*. http://wikileaks.wikimee. org/cable/2009/01/09BRATISLAVA9.html. Accessed 8 Jan 2009.

Embassy Cairo. (2006). Senior advisor for agricultural biotechnology advocates science-based regulatory framework in Egypt and Middle East. *WikiLeaks Public Library of US Diplomacy*. https://wikileaks.org/plusd/cables/06CAIRO2165_a.html. Accessed 10 Apr 2006.

Epstein, S. (1996). *Impure science: AIDS, activism, and the politics of knowledge*. Berkeley: University of California Press.

European Commission. (2012). *Innovating for sustainable growth: A bioeconomy for Europe. 13.2.2012 COM(2012) 60 final*. Brussels: European Commission.

Gieryn, T. F. (1983). Boundary-aork and the demarcation of science from non-science: Strains and interests in professional ideologies of scientists. *American Sociological Review, 48*(6), 781–795.

Goven, J., & Pavone, V. (2015). The bioeconomy as political project: A Polanyian analysis. *Science, Technology and Human Values, 40*(3), 302–337.

Hecht, G. (1994). Political designs: Nuclear reactors and national policy in postwar France. *Technology and Culture, 35*(4), 657.

Hilgartner, S. (2007). Making the bioeconomy measurable: Politics of an emerging anticipatory machinery. *BioSocieties, 2*(3), 382–386.

Hurlbut, J. B. (2018). Laws of containment: Control without limits in the new biology. In I. Braverman (Ed.), *Gene editing, law, and the environment: Life beyond the human* (pp. 77–94). New York: Routledge.

Institute for Fisheries Resources et al., & V. S. M. Burwell et al. (2016). Case 3:2016cv01574, March 30, 2016, California Northern District Court. San Francisco: San Francisco Office. https://dockets.justia.com/docket/california/candce/3:201 6cv01574/297176.

Jasanoff, S. (1990). *The fifth branch: Science advisers as policymakers*. Cambridge: Harvard University Press.

Jasanoff, S. (2005). *Designs on nature: Science and democracy in Europe and the United States*. Princeton: Princeton University Press.

Jasanoff, S. (2012). Taking life: Private rights in public nature. In K. S. Rajan (Ed.), *Lively capital: Biotechnologies, ethics, and governance in global markets* (pp. 155–183). Durham: Duke University Press.

Jasanoff, S., & Kim, S.-H. (2015). *Dreamscapes of modernity: Sociotechnical imaginaries and the fabrication of power*. Chicago: University of Chicago Press.

Juma, C. (2016). *Innovation and its enemies: Why people resist new technologies*. New York: Oxford University Press.

Letter to President Obama, Roberts, R. J., Van Eenennaam, A., Juma, C., Walton, M., Beachy, R., Carlson, D., Giddings, E. V., & Pommer, J., et al. (2014). Scientist executive letter to President Obama on biotechnology. www.ftrw.org/scientist_executive_letter_to_president_obama_on_biotechnology.docx. Accessed 17 Sept 2014.

Miller, C. (2015). Globalizing security: Science and the transformation of contemporary political imagination. In S. Jasanoff & S.-H. Kim (Eds.), *Dreamscapes of modernity: Sociotechnical imaginaries and the fabrication of power* (pp. 277–299). Chicago: University of Chicago Press.

Obama, B. (2015). United States, State of the Union Address. Remarks by the President in State of the Union Address/January 20, 2015. https://obamawhitehouse.archives.gov/the-press-office/2015/01/20/remarks-president-state-union-address-january-20-2015.

Obama, B. (2016). United States, State of the Union Address. Remarks of President Barack Obama-State of the Union Address as delivered/January 13, 2016. https://obamawhitehouse.archives.gov/the-press-office/2016/01/12/remarks-president-barack-obama-%E2%80%93-prepared-delivery-state-union-address.

OECD. (2009). Bioeconomy to 2030: Designing a policy agenda. http://www.oecd.org/futures/long-termtechnologicalsocietalchallenges/42837897.pdf.

Rajan, K. S. (2006). *Biocapital: The constitution of postgenomic life*. Durham: Duke University Press.

Smith-Spark, L., & Hanna, J. (22 Apr 2017). March for science: Protesters gather worldwide to support "evidence." *CNN*. https://www.cnn.com/2017/04/22/health/global-march-for-science/index.html.

Specter, M. (2009). *Denialism: How irrational thinking hinders scientific progress, harms the planet, and threatens our lives*. London: Penguin Press.

Staffas, L., Gustavsson, M., & McCormick, K. (2013). Strategies and policies for the bioeconomy and bio-based economy: An analysis of official national approaches. *Sustainability, 5*(6), 2751–2769.

Towers, L. (2016). AquaBounty starts AquAdvantage Salmon trials in Brazil, Argentina. https://thefishsite.com/articles/aquabounty-starts-aquadvantage-salmon-trials-in-brazil-argentina. Accessed 4 Feb 2018.

United States Food and Drug Administration. (2012). AquAdvantage Salmon draft environmental assessment. https://www.fda.gov/downloads/AnimalVeterinary/DevelopmentApprovalProcess/GeneticEngineering/GeneticallyEngineeredAnimals/UCM333102.pdf. Accessed 4 May 2012.

United States Food and Drug Administration. (2015). AquAdvantage Salmon-Response to public comments on the environmental assessment. http://www.fda.gov/AnimalVeterinary/DevelopmentApprovalProcess/GeneticEngineering/GeneticallyEngineeredAnimals/ucm280853.htm. Accessed 7 Aug 2016.

United States Secretary of State. (2007). FY 2008 biotechnology outreach strategy and department resources. *WikiLeaks Public Library of US Diplomacy*. https://wikileaks.org/plusd/cables/07STATE160639_a.html. Accessed 27 Nov 2007.

House, White. (2012). National bioeconomy blueprint, April 2012. *Industrial Biotechnology, 8*(3), 97–102.

Tess Doezema is a Doctoral Candidate in the Human and Social Dimensions of Science and Technology in the School for the Future of Innovation in Society at Arizona State University, and a Visiting Fellow with the Program on Science, Technology & Society (STS) at the Harvard Kennedy School of Government.

The Institutionalization of an Envisioned Future. Sensemaking and Field Formation in the Case of "Industrie 4.0" in Germany

Uli Meyer

Abstract

Technology-based, envisioned futures have a significant influence on the dynamics of technological development and, as a consequence, on societies. If such envisioned futures are successful, they contribute to the formation of issue-based fields with the envisioned future at its core. Within such a field, organizations orient and coordinate their activities in pursuit of the envisioned future. This article uses the example of "Industrie 4.0" in Germany to analyze why, how and under what circumstances such imagined futures tend to emerge, diffuse and stabilize. In particular, it highlights the early phases of envisioned technological futures before they are widely known and accepted. The paper brings together concepts from organization studies (OS) as well as from science and technology studies (STS), the most prominent of which being "expectations in technological developments" (van Lente 2000; van Lente and Rip 1998a). For a more detailed understanding of an envisioned future's impact on the present, I argue that it is essential to analyze the role and the activities of organizations and the formation of *organizational fields* in such processes. I show how Weick's concept of *sensemaking* and the related ideas of *enactment* and *sensegiving* can contribute to this idea of *field formation*. This combination enables a better understanding of the role of organizations, especially in the early phase of such processes. I argue that one main virtue of an envisioned future, when successful, is its ability to provide orientation to a multitude of different organizations.

U. Meyer (✉)
Munich Center for Technology in Society, Technical University of Munich,
Munich, Germany
e-mail: uli.meyer@tum.de

© Springer Fachmedien Wiesbaden GmbH, part of Springer Nature 2019   111
A. Lösch et al. (eds.), *Socio-Technical Futures Shaping the Present*,
Technikzukünfte, Wissenschaft und Gesellschaft / Futures of Technology,
Science and Society, https://doi.org/10.1007/978-3-658-27155-8_6

1   Introduction

In 2017 the conservative German Konrad Adenauer Foundation and the "Stiftung Neue Verantwortung" (SNV), an economically liberal think tank, warned that the idea of "Industrie 4.0" had risen to such prominence in Germany so as to eclipse all other aspects of the digital transformation not connected to industrial production (Lorenz 2017). "Industrie 4.0" has become the reference point for many economic and political activities in Germany—with an influence that extends well beyond the industrial sphere. In this paper, I use the success of this catchword as an example of an envisioned future with its own inherent dynamic and a clear potential to shape the present. My focus in the following lies on the early phases of an envisioned future and the road to widespread acceptance. I argue that actors use such envisioned futures as a source of orientation for their actions. They are an universal tool for sensemaking, as well as for justifying activities and creating legitimacy. While these visions supply a realistic and adequate description of a possible technological future, this is not the primary feature responsible for their success. Historically, we have seen no shortage of possible futures, but few are remembered and even fewer succeed. So, why and how do envisioned futures become successful?[1]

2   The Term "Industrie 4.0"

The term *Industrie 4.0* describes the use of cyber-physical systems in industry settings. It was promoted in Germany by a group of actors that included industry associations, the German Academy of Science and Engineering (acatech), policy makers (e.g. the German Federal Ministry of Education and Research, BMBF), and labor unions. The term was first used in a 2011 paper and mentioned for the first time in public at the international trade fair Hannover Messe in 2012 (Kagermann et al. 2011; Pfeiffer 2015). Since then, it has enjoyed a steep rise and developed into a powerful and influential term. In 2016, German Chancellor Angela Merkel and US President Barack Obama officially opened the Hannover Messe. Four years after the term had been introduced at that same venue, one of the most

---

[1]Drawing on the model presented in this paper, one path for future research could be to compare successful examples like "Industrie 4.0" with other less successful terms that failed to resonate on a large scale. But it should come as no surprise that finding those ideas that never "made it big" poses some methodological problems.

well-attended forums at the 2016 event was "Industrie 4.0 Meets the Industrial Internet".[2] "Industrie 4.0" had become 'the' topic of the fair. The event focused on the digital transformation of industries, and the United States was the official partner country. This was reason enough for Barak Obama to make the Hannover Messe the destination of his last official visit to Germany as President of the United States. Several countries have "imported" the German term, especially in the Asian region, but also a number of Eastern European countries. Just five years after the term was coined, a variety of organizations and programs in support of Industrie 4.0 had already sprung up.

Numerous definitions of Industrie 4.0 exist, which is a property of the envisioned future I will come back to later in this article. One of the more frequently cited definitions stems from "Plattform Industrie 4.0", which is one of the newly founded organizations:

> "Industrie 4.0 combines production methods with state-of-the-art information and communication technology. The driving force behind this development is the rapidly increasing digitization of the economy and society. It is changing the future of manufacturing and work in Germany: In the tradition of the steam engine, the production line, electronics and IT, smart factories are now determining the fourth industrial revolution." (Industrie 4.0 plattform[3])

Besides being one of the more popular definitions, this one has the advantage of presenting in a condensed form most of the crucial properties of this particular envisioned future. It highlights (a) technological foundation of the term, (b) the future orientation of the term, (c) its impact on industry but also on society, and (d) it explains the term as being part of a larger historical development. The general idea is that industrial production will reach a new level based on the integration of cyber-physical systems (Hirsch-Kreinsen 2016). The consequences of this "revolution" are described as disruptive for the economy as well as for society as a whole. What is more, and this is an important aspect of the term, if Germany and Europe pour all of their efforts into achieving this vision, they will benefit significantly—both economically and socially—from this transformation. The "Plattform Industrie 4.0", which is the source of this definition, also reflects the

---

[2]Known as "Industrie 4.0" in Germany, this phenomenon is referred to as "industrial internet" in the US, and other countries use other terms. In France, for example, the term "l'Industrie du Future" (the industry of the future) is used.

[3]http://www.plattform-i40.de/I40/Navigation/EN/Industrie40/WhatIsIndustrie40/what-is-industrie40.html, accessed June 22, 2016.

success of the envisioned future behind Industrie 4.0. It is a meta organization (Ahrne and Brunsson 2005) whose members are companies, industrial associations and political actors.[4] Other indicators of this vision's success are the special funding programs which have been created to accelerate the implementation of Industrie 4.0 technologies. For example, the German Federal Ministry of Education and Research (BMBF) has a special program for the "Zukunftsprojekt Industrie 4.0" ("future project Industrie 4.0").[5]

This article argues that the envisioned future of an "Industrie 4.0" has become so successful and influential due to the emergence and formation of an organizational field (DiMaggio and Powell 1983; Hoffman 1999) around this specific issue. This field, in turn, is the result of the individual sensemaking activities of many different organizations.

3   Concepts of Socio-Technical Futures

Some examples of successful envisioned futures from recent decades are *Moore's Law*, *high-definition television* (HDTV), and the *information superhighway*. There is a variety of concepts in science and technology studies (STS) and the sociology of technology to describe how such stories, ideas and imaginations about future technologies impact the present.

The concept of *imaginaries* can be applied to analyze the relationship between national policies and political institutions on the one side and science and technology (Jasanoff and Kim 2009) on the other.[6] This perspective clearly shows how ideas about the future are always political and rooted in existing structures (Felt 2015). *Visions of technology or guiding visions* (in German: *Leitbilder*) (Dierkes et al.1996) is a concept that examines how shared understandings and ideas about technical futures function as a cognitive framework for decision-making under uncertain conditions, also within organizations (Dierkes 1988, p. 54). The concept of *organizing visions* (Swanson and Ramiller 1997) asks how and why organizations adapt different information technologies (IT) in specific ways. Most relevant for this paper are van Lente and Rip's writings on *expectations in technological*

[4]http://www.plattform-i40.de/I40/Navigation/EN/ThePlatform/PlattformIndustrie40/platt-form-industrie-40.html, accessed June 22, 2016.

[5]https://www.bmbf.de/de/zukunftsprojekt-industrie-4-0-848.html, accessed June 22, 2016.

[6]The authors draw attention to contrasting national imaginaries, such as "atoms for peace" in the US and "atoms for development" in South Korea (Jasanoff and Kim 2009).

developments (van Lente and Rip 1998a). This perspective describes the dynamics of technological futures, which often start as an *option* and then, through stories, become a *promise*, a *requirement* and finally a *necessity*.

This perspective most closely approximates the idea of envisioned futures presented here. Expectations about the future of technology, van Lente and Rip argue, reduce uncertainty by providing scripts that can be used in action. Through such scripts, "expectations allocate roles for selves, others, and (future) artifacts" (van Lente and Rip 1998a, p. 203). They also connect previously unconnected actors. Scripts often take the form of "statements, brief stories or scenarios" (van Lente and Rip 1998a, p. 205). The authors show how expectations in technological development do not need to be grounded in technological breakthroughs but often are based on "rhetorical invention" (Van Lente and Rip 1998b, p. 225). Van Lente and Rip illustrate how expectations transform descriptive technological scenarios into requirements and ultimately necessities. "Opportunities (generated within—or without—'the protected space') presented as promises get accepted and become part of an agenda; and are subsequently converted into requirements that guide the search processes" (van Lente and Rip 1998a, p. 223). Existing technologies, in this case, become problematic in the sense that they cannot live up to the promises of new technologies. "Once technical promises are shared, they demand action, and it appears necessary for technologists to develop them, and for others to support them" (van Lente and Rip 1998a, p. 216). Organizations often participate in the development of technologies described in technological futures because they do not want to fall behind. But by doing so they actually create the same dynamic they fear, because falling out of touch is, to a certain extent, inevitable: "As soon as expectations are shared they assume a life of their own" (van Lente and Rip 1998a, p. 217).

This paper takes the above perspective as a starting point and delves into the question of why organizations subscribe to envisioned futures and the technical options they entail—especially in the earliest stages, that is, before an option has become successful or been transformed into a necessity. While van Lente and Rip highlight the properties of promises that are already out in the open ("once [...] they are shared" and "get accepted"), this paper scrutines the emergence of envisioned futures or the processes that support these visions prior to that point. How do envisioned futures come to be shared by many and assume trappings of inevitability? What happens for an envisioned future to achieve that level of success and what leads to its diffusion?

The second relevant deviation between my perspective and the "expectation perspective" is that the latter assumes that the credibility of an expected technological future is the crucial dimension for its success. The authors then analyze how this credibility depends on a technology's current performance, the plausibility of the path forward and the end target, which is constructed (Bakker et al. 2012, p. 1060). I argue that credibility is not as crucial as depicted by these authors. For the envisioned future of Industrie 4.0, for example, the historical development, present state and future path of this option are described only in very vague and general terms. These broad strokes allow very different actors to connect this narrative with their own situation. More relevant than providing a highly credible description, a general narrative does succeed in effectively reducing uncertainty. It does so by creating an impression of urgency, wich is connected to existing debates, and at the same time providing a solution for this problem.

3.1 Futures Concepts in STS, a Preliminary Summary

The different approaches in STS described above share some common properties. First and foremost, they all describe how ideas about technological futures influence the present. These futures are (a) imagined in the present and based on technology and predictions about its future development. They represent (b) widely shared expectations about the future and (c) are depicted as possible, attainable and desirable (or worthy of pursuit). They (d) help to reduce uncertainty by providing orientation, coordination, and motivation, and (e) are phrased in a meaningful and memorable way. This description provides invaluable insights for the phenomenon of envisioned futures: Envisioned futures are always political, and they exist and are manifested on very different levels. They develop in a larger historical context, including existing societal frames and ideas about technology. Such envisioned futures change their dynamics considerably as they diffuse and stabilize. They also bring together very different actors with very different ideas about those futures.

As described above, van Lente and Rip emphasize the inherent dynamics of technological futures and how they turn from a promise—or a promising solution—into a veritable mandate. My goal in this article is to build on this line of research by providing a detailed analysis of how an envisioned future develops in individual organizations and eventually spurs the formation of an entire organizational field. To use "Industrie 4.0" as an example: The term was coined by industry associations, policy makers and unions, and then picked up very quickly by other actors. Meanwhile, the existence and importance of "Industrie 4.0" is taken for granted by most actors within the field. Most of the crucial actors in this field are organizations, that is, companies, state agencies, NPOs, labor unions, industry associations, and so on. So what motivated all of these different organizations to sign up for the same envisioned future? And how did this envisioned future diffuse to encompass several different organizations? To address this question, it is important to understand why organizations refer to and use visions.

Much has already been written about how social dynamics become self-reinforcing or at least self-stabilizing (Arthur 1989; Meyer 2013) once they reach a certain threshold (Granovetter 1978) or "tipping point" (Gladwell 2000). Social dynamics tend to gain momentum because they already have it. This has been shown for research (Fujimura 1988), for technology and economic processes (Arthur 1988), for organizational fields in general (DiMaggio and Powell 1983) and for many other processes in particular. But now, research needs more answers about what happens at the beginning of these proceeses. In other words, why do different actors get involved in a vision before the scales tip in its favor? One thing is clear: the claim that powerful actors are the driving force behind these visions does not go far enough. This dynamic can only be properly understood—at least that is the argument of this article—by considering the role of organizations in these processes. So, my objective in the following will be to show what organizations do to create, influence and change the dynamics of an envisioned future.

4   Envisioned Futures and Sensemaking

Karl Weick's sensemaking perspective provides a useful starting point for explaining why and how envisioned futures start to emerge. The sensemaking perspective analyzes phenomena at the level of individual organizations.[7] This allows us to obtain detailed insights into why different organizations (or categories thereof) subscribe to a specific story of technological evolution. In addition, ideas have already been developed about how to link this perspective to a more institutional point of view.

---

[7]Sensemaking can also be described as a micro-level theory: originally, sensemaking described processes involving individual actors, which were then transferred to an organizational level. More generally, sensemaking does not represent a consistent theoretical approach. Very different offshoots of this perspective have been developed in recent decades (Maitlis and Christianson 2014). Therefore, what I present here is the summary of one specific perspective within the sensemaking literature.

"It is this institutionalizing of social constructions into the way things are done, and the transmission of these products, that links ideas about sensemaking with those of institutional theory. Sensemaking is the feedstock for institutionalization" (Weick 1995, p. 87)

While not the most elaborated part of his theory, Weick also gives hints as to how sensemaking on an individual and organizational level leads to institutionalization and broader (societal) dynamics. Being able to theoretically link these levels is necessary in order to describe how sensemaking at the individual organizational level can develop into an envisioned future that is widely accepted and shared as described by van Lente and Rip.

4.1 Sensemaking

True to its name, sensemaking describes how actors—both individual and collective—make sense of events and the world at large. This perspective's basic assumption is that the world around us and different problems do not present themselves as given. Actors must construct them from the materials at their disposal. This is especially the case in puzzling, troubling, and uncertain situations. Actors select what they treat as a situation and the elements that constitute it. Sensemaking is a cyclic and iterative sequence:

"The cycle begins as individuals form unconscious and conscious anticipations and assumptions, which serve as predictions about future events. Subsequently, individuals experience events that may be discrepant from predictions. Discrepant events, or surprises, trigger a need for explanation, or post-diction, and, correspondingly, for a process through which interpretations of discrepancies are developed" (Weick 1995, p. 5).

Sensemaking takes place in the (verbal) exchange between actors. It is a quintessentially social process (Weick et al. 2005, p. 409) and the process through which the social world is constructed and interpreted (Gephart 1993, p. 1485). As a process of social construction, sensemaking is very much in line with the work of Berger and Luckmann (1967). Sensemaking activities allow actors to create a collective understanding of the world and a foundation for collective action (Sandelands and Stablein 1987; Starbuck and Milliken 1988; Isabella 1990; Sackman 1991; Weick and Roberts 1993).

The important aspect of sensemaking is its fundamental nature. The process is more far-reaching than a mere interpretation of the situation at hand: "Most descriptions of interpretation focus on some kind of text. What sensemaking does is address how the text is constructed as well as how it is read. Sensemaking is about authoring as well as reading" (Weick 1995, p. 7). Or, in other words: "Sensemaking is not about the discovery of a given situation, it is about the invention of the situation" (Weick 1995, p. 14). This "authoring" aspect in sensemaking processes also means that situations will be defined very differently by different actors. The specific form and the outcome of sensemaking heavily depends on the perspective, experiences, and (organizational) context of the sensemaking actor. This is especially true for organizations. They supply a toolkit for sensemaking that "include[s] the standards and rules for perceiving, interpreting, believing, and acting that are typically used in a given cultural setting" (Sackman 1991, p. 33).

Sensemaking is rooted in uncertainty. This represents another overlap with the concept of envisioned futures and their uncertainty-reducing properties. Sensemaking specifies the kind of uncertainty which it provides a remedy for, which is the same that makes envisioned technological futures so helpful: This uncertainty does not exist for a lack of information. Rather, it comes from the lack of a clear interpretation of the information that is available. Successful future visions provide an interpretative framework for existing information. Thus, they are not about detailed facts concerning future technology. Instead, they provide interpretative frames that help people make sense of what these technologies mean, what they stand for, and what relevance they have for different groups.

Sensemaking—the construction and explanation of a situation—often takes place through language (Taylor and Van Every 2000). In many cases, talk or narrating are its vehicles (Balogun and Johnson 2004; Currie and Brown 2003; Dunford and Jones 2000; Gabriel 2004). Language is "the preferred sensemaking currency" (Boje 1991, p. 106). Moreover, narratives are especially powerful when it comes to constructing and communicating explanations. Narratives help to construct a plausible story, which also includes rationalizing and evaluative elements (Brown et al. 2008; Currie and Brown 2003; Humphreys and Brown 2002). This provides a useful link to the explanation of envisioned futures presented here. Successful envisioned futures contain a narrative that allows actors to explain their world and their own role in it.

Sensemaking can be the source of dynamic processes beyond the control of individual organizations. If an envisioned future like the one described by "Industrie 4.0" is successful, it is necessary to understand how this concept became part of the sensemaking process of different organizations. Moreover, how can this envisioned future contribute to the reduction of uncertainty? And, how does one narrative accomplish this for a wide variety of organizations?

4.2 Enactment, Sensegiving and Role Requirements

The consequences and impacts of sensemaking are not limited to the individual organization. To the contrary: Sensemaking is the foundation for the dynamics which lead to the formation of fields or, more specifically in this case, to *envisioned future fields*. Based on their sensemaking activities, organizations act and interact with their environment. They *enact* their ideas about the situation in which they find themselves: "In organizational life, people often produce part of the environment they face" (Weick 1995, p. 30). By making sense of their situation in a specific way and acting accordingly, people influence and change their environment. This relationship between sensemaking and action can have iterative effects. Actors may stabilize their world and worldview by enacting their own perspective on how the world works. Of course, they can and do also fail dramatically, but often, sensemaking reinforces the situation, and vice versa, triggering dynamics very similar to self-fulfilling prophecies. In this case, action rooted in sensemaking shapes the world that has been constructed by those same sensemaking processes.

Furthermore, enacted ideas and understandings change the world for others. Large parts of an organization's environment consist of other organizations. In consequence, organizational sensemaking and the enactment of an organization depends largely on the sensemaking and enactment of other organizations. For example, by referring to Industrie 4.0 as a probable and promising future, organizations also make others aware of that envisioned future. They demonstrate its relevance for their own strategic planning. These may be the unintended side effects of intraorganizational sensemaking processes.

It should come as no surprise that organizations actively try to influence the sensemaking activities of other organizations. Such *sensegiving* activities (Gioia and Chittipeddi 1991) attempt to influence other organizations' sensemaking processes. They aim to create meanings for a target audience (Gioia and Chittipeddi 1991). This is a very important element that links organizational actors to higher-level dynamics. Actors not only use an envisioned future in their own sensemaking, but they also influence their environment in different ways as they attempt to actively spread that future. The results can be very similar to the dynamics described by van Lente: descriptions of a possible future turn into promises and ultimately into inevitabilities.

Every actor engages in sensemaking. For some actors, however, they are even more relevant than for others. This is often because the performance of explicit forms of sensemaking and sensegiving make up (at least part of) their role and the expectations tied to it. The actors with the most at stake in an envisioned future, such as companies and associations, are a case in point, as they are required to engage in active and explicit sensemaking activities. Beyond the basic necessity shared by all actors to make sense of their world, these group of actors are expected to do this in a specific way. Managers and politicians have to fulfill certain expectations based on sensemaking and sensegiving. They must act strategically, plan, learn from the past, and find solutions for present and future problems. Moreover, they are expected to communicate all of these activities and make them visible, for example, to different stakeholders within and outside their own organizations. These actors are required to explain and justify their actions, not only to themselves, but also when addressing others like stakeholders or members.

The dynamic potential found in sensemaking, enactment, and sensegiving is a very important stepping stone to institutionalization. If organizations are successful in their sensegiving activities, that content can become an "externally specified objective reality" (Weick 1995, p. 36). At some point, both external actors and those participating in the sensemaking process perceive an envisioned future as fact. Sensemaking also has the potential to connect actors without prior ties. This property was described above as one of the aspects of envisioned futures found in STS concepts. The sensemaking perspective, I argue, provides a more precise description of such processes. Moreover, in doing so, it supplies a crucial mechanism for the diffusion of ideas and values and builds the foundation for field-level institutionalization.

5   Envisioned Futures and the Formation of Fields

Through activities like sensemaking, enactment, and sensegiving, new connections are created between different actors. Organizations develop a sense of which other organizations are relevant for an envisioned future; there is an increase in the flow of information as well as interactions, and patterns of dominance and coalition emerge. In other words, an organizational field starts to form (DiMaggio and Powell 1983). Organizational fields are specific kinds of meso orders (Fligstein and McAdam 2012). They consist of the network of actors—mainly organizations (DiMaggio and Powell 1983)—their relational network, and the norms and values which develop within this field. Here, it becomes apparent that this concept is deeply rooted in institutional theory. Fields are not only characterized by interaction patterns but also by institutionalization processes. Or, to borrow a phrase from DiMaggio and Powell, fields are the source of "institutional life" (1983).[8]

Fields form around a specific issue "[...] that becomes important to the interests and objectives of a specified collective of organizations. Issues define what the field is, making links that may not have previously been present." (Hoffman 1999, p. 352). An envisioned future can become such an issue, based on which new and/or specific relations emerge between actors.

Also, and in accordance with Hoffman, the internal structure of a field consists of different populations. A population contains all organizations that share similar goals and similar perspectives towards the envisioned future. Yet it is important to note that if organizations form around an issue, this does not imply that all of those organizations or populations share the same perspective on that issue. On the contrary, fields become hotbeds for "debates in which competing interests negotiate over issue interpretation" (Hoffman 1999, p. 351). Even if all organizations in the field do share an interest in the issue, their perspective on the issue, how they frame it and their interests, may vary to a large extent.[9]

Deviating from the conventional (organizational) field perspective outlined above, I explain the emergence of fields surrounding envisioned futures by drawing on sensemaking processes and their emergent properties. Sensemaking processes lead to the formation of interorganizational networks, the first step in the formation of fields. If, on the other hand, institutionalization has progressed to the point that the issue comprising the field has become taken for granted, the field itself will heavily influence sensemaking processes. In this case, an envisioned future does not require any explanation or justification. On the contrary, it can be used to explain and justify other activities.

6   Sensemaking and Envisioned Futures

Combining sensemaking with the idea of institutionalization and organizational fields also allows for a process perspective. Starting with the first organizations that subscribe to a newly envisioned future, the researcher can trace and analyze

---

[8]Therefore, in contrast to other concepts such as "industries," organizational fields are not defined by a set of similar organizations, but by the emergence of a social structure linking the involved actors.

[9]Hoffman illustrates this for the case of environmentalism in the United States. Environmental NGOs and chemical companies were involved with this issue. These two groups shared an interest in the issue, but little else.

the development of that vision until it is stabilized within a field and becomes widely accepted and "shared by many," in the words of van Lente and Rip.

The starting point for this model is the dynamic analysis of how such an envisioned future develops between heterogeneous actor groups. Naming, narrating and describing an envisioned future are important early steps.[10] In rather vague terms, such a vision will bring together ideas about future technological development and references to societal dynamics and link these two elements to an anticipated (social) transformation. Actors use these elements in sensegiving processes to publicize "their" envisioned future and try to convince others of its relevance and importance. Paramount to the success of an envisioned future is a memorable name that allows for a variety of interpretations. This "catchy" term must permit multiple and vivid imaginations of a positive future. Such a label is necessary in order to mobilize many and different actors. Envisioned futures are political from the moment they are born. There is no such thing as a neutral, interest-free vision of the future. Sensegiving, for example, is always an attempt to make one's own perspective the dominant one. Different actors and actor groups can profit from the future they champion in different ways.

Most envisioned futures never make it past this early phase. However, if they continue to diffuse and become more widely recognized, it is because many actors consider them useful, and use them, in their sensemaking and sensegiving activities. An envisioned future will continue its "career" as long as more and more actors refer to it. They use it to make sense of their situation, explain their activities, and "help others out" (i.e. sensegiving) by supplying their own preferred and ready-made interpretations. As they acquire momentum, envisioned futures permit different actors to reduce uncertainty. They provide guidance for the present by describing a possible future and a plan to get there—not by furnishing additional information, but by supplying an interpretative frame for existing information. Successful envisioned futures address, bring together and connect many different groups of actors: technology manufacturers, (potential) users, political actors and so on, and they must be relevant for all of these groups. An envisioned future is valuable because it provides a point of reference for decisions and action. As many different actors start to refer to the same envisioned future, their modes of reference vary (Besio and Meyer 2015), resulting in multiple interpretations of the envisioned future.

---

[10]These steps already include a variety of sensemaking processes, but they do not constitute the focus of the described model. Nor do they represent the very first steps. This type of storyline already builds on pre-existing elements, ideas, and developments.

Given all of these properties, it is of course better when an envisioned future is also relatively realistic. But real attainability is not the basic mechanism that drives envisioned futures. Here, the model diverges from what Bakker et al. (2012) describe. Without a doubt, some level of credibility is necessary for an envisioned future to gain sway and survive; highly unrealistic claims are not likely to succeed. But the specific quality of an envisioned future is its ambiguity. The quintessential envisioned future presents itself as a solution for many different problems: legitimacy issues, technological or business decisions, uncertainty, and ambivalence.

The stories and narratives that drive an envisioned future also connect the specific technological aspects it contains to larger societal issues and debates. For technology-based envisioned futures, these links pertain, first and foremost, to ideas about progress and innovation in society. In addition, a narrative can also link more specific concepts and ideas to the technology. This is very much in line with how van Lente and Rip describe technological futures. Technological futures are always connected to ideas of progress and technological evolution. They also become linked to other technologies, many from the past, "creating trajectories of technologies from past to present to the future technology, which then in such a context seems natural and unavoidable" (van Lente 2000, p. 46).[11]

As such links continue to be forged, they trigger a cascade of sensemaking and sensegiving activities. Bonds emerge between "previously unconnected actors" (van Lente and Rip 1998a, p. 205), who start to interact more and more; interorganizational structures develop and information flows and mutual awareness increase (DiMaggio and Powell 1983). In other words, a field starts to emerge and, at some point, a tipping point is reached and the envisioned future becomes self-stabilizing based on its own content and dynamics. Once the envisioned future has acquired its own momentum, the underlying mechanisms shift, since it has become taken for granted within the field. Actors refer to the envisioned future out of habit, or because "everybody's doing it". Once a vision, this future has successfully undergone the transformation from an option to a promise and, finally, a requirement. Two more elements contribute to this development. First, existing organizations not only adapt their activities, but also their structures to the envisioned future. Positions are created, projects and groups formed, maybe even new departments. Second, new organizations emerge. The most obvious

---

[11]This also means that fields addressing envisioned futures can develop very differently in different countries. This is very much in line with the imaginaries perspective described earlier.

cases are organizations explicitly founded based on the envisioned future. The previously mentioned "Plattform Industrie 4.0" is a point in case.

A successful envisioned future is characterized by self-reinforcing dynamics, and possibly path dependencies (Arthur 1989; Meyer 2016). Even if created intentionally, it cannot necessarily be controlled and steered in line with its creators' intentions. Numerous different sensemaking processes may lead in sum to very different outcomes and developments than originally planned by any group. The vagueness and ambiguity of envisioned futures contribute even more to these effects.

In the following, the model outlined here is used to analyze the success of Industrie 4.0. The starting point for the explanations supplied below is the role and activities of different kinds of organizations—something that is normally missing in models of socio-technical futures.

7 "Industrie 4.0" as an Envisioned Future

This study on the development of the envisioned future expressed by the term "Industrie 4.0" is based on over 160 interviews we conducted between November 2015 and December 2017. Most of the interviews lasted between one and two hours, with some deviations.[12] We conducted the interviews with managers, engineers, researchers and representatives of a variety of associations, ministries, labor unions, and workers' councils, all of whom were concerned with Industrie 4.0 in one way or another.[13] Research data also included a variety of different publications such as press releases, scientific articles, conference proceedings and so on. The transcribed interviews and additional materials were coded and analyzed using QDA software.

As described in the introduction, "Industrie 4.0" was coined in 2011. The term and its corresponding narrative have a number of properties that contribute to its success. Industrie 4.0 combines the idea of a transformation in the industrial sector with the idea of software development and digitization. It also introduces the idea of versioning (2.0, 4.0) into an industrial context. The related narrative also links these different dimensions. Though it might seem boring when compared to overly hyped terms like the *sharing economy* or *big data*, this story of change

---

[12]Some of the interviews lasted only 15 min; the longest lasted over four hours.

[13]The fact that many actors are required to perform active sensemaking as part of their job is also useful for data collection. Most were quite willing to do an interview and most have also made public statements about their stance towards Industrie 4.0.

and transformation tells us that we are in the midst of an industrial revolution. However, this shift is also part of a larger transformation, because it is already the fourth revolution of its kind. So, in addition to creating a link between industry and software, this story supplies a larger historical narrative of technological and industrial dynamics and success. The fourth industrial revolution is the logical and inevitable continuation of a development that started with the steam engine, continued with the production line, then progressed to electronics and IT, and has now reached the stage of the smart factory. This story of technological and societal development is narrated in a specific context, which it references and integrates. An important element of this context is society's ubiquitous focus on innovation. In today's society, innovation is presumed to be the cure-all for most, if not all, problems. This basic orientation has been termed an "innovation imperative" (Windeler et al. 2018). Besides a general orientation toward innovation and progress—something all envisioned futures share—Industrie 4.0 addresses more specific topics and one of those is how Germany can stay competitive in a global market. The question of how Germany can remain a high-wage manufacturing country has been discussed for at least a decade. Industrie 4.0 presents itself a solution:

> "And especially for production, digitization is an urgent matter, because Germany is still a region with a high level of production. We still have a strong, traditional industrial manufacturing, something you will not find in other European countries or the US". (interview, industry association representative)

This perceived need for change, in turn, is fueled by a widely shared impression that Germany has already lost its competitive edge in many IT-related areas such as software, the internet, the platform economy and so on. Industrie 4.0 builds on this perception and presents a plan to prevent this from happening in manufacturing as well. This need applies to nearly all areas of society, but it is especially relevant for the industrial and political domains. Industrial sectors need to stay competitive on both an organizational and a national level. And achieving this need is depicted as a permanent, never-ending struggle.

Industrie 4.0 promises a better future and highlights the urgency of prompt action to achieve the envisioned objectives. If people act now, Industrie 4.0 will not only bring improvements for the economy, but also for society as a whole—that's the promise. And if companies—and entire nations—want to be successful in the future, they have to prepare now. Industrie 4.0 tells a story which unites and supports numerous different actors in their sensemaking process. That story is specific in some respects (e.g. new technologies such as cyber-physical systems, data integration, automation, and the smart factory will usher in the next industrial revolution), but open enough (e.g. what exactly is a smart factory or what is so new about automation?) to be used by different actors.

The first group actively promoting Industrie 4.0 has been industry associations. This is one group in contemporary society that is called on to interpret events and developments and actively share those interpretations with a broader audience. Industry associations are expected to provide orientation and support to their members and Industrie 4.0 is one attempt to do so. Right from the start, industry associations have engaged in sensemaking activities related to Industries 4.0, which is deemed relevant for their member organizations. When it comes to those members, industry associations play an active role in getting them on board. One industry association representative described these sensegiving activities in an interview:

> "And we have to explain to a lot of people why this is useful. What do I need 'Industrie 4.0' for? For example, I am a screw manufacturer, so why do I need Industrie 4.0? We need to be able to answer those types of questions." (interview, industry association representative)

This way not only member organizations (or companies) have been addressed, but also at political actors. The story of Industrie 4.0 allowed different associations to showcase their activities and demonstrate their relevance: first, by providing this envisioned future; and second, by ensuring their relevance within that future. This strategy aimed to garner support for Industrie 4.0 among member organizations. In this early stage, there was still no field. Only a few people knew about the term Industrie 4.0 and of those few, most were skeptical.

One of the strongest proponents of Industrie 4.0 is and has been Bitkom, Germany's association for the IT industry. For Bitkom, Industry 4.0 presents a promising opportunity to extend the association's sphere of influence. The introduction of cyber-physical systems to a wide range of industrial companies would imply and increase influence of computer science in these companies. Bitkom states this very clearly on its website:

> "Never before has one sector been as important for technology driven economic development as the IT sector is for the topic of Industrie 4.0" (Bitkom website)[14]

---

[14]https://www.bitkom.org/Themen/Branchen/Industrie-40/Vision-Industrie-40.html, accessed: 08.08.2016.

The second group that has been actively involved in the early stages and propagation of this envisioned future is political actors. For this group, Industrie 4.0 provides a solid—yet not too specific—foundation for political programs. It can be—and is—the basis for very different kinds of action. One way that Industrie 4.0 is used in sensemaking and sensegiving in this context is as a promise for reindustrialization or as a way to stay competitive as a region or a nation. A variety of programs, funding schemes, and initiatives has been created with the aim of preparing the country for the fourth industrial revolution. Different federal ministries as well as some regional governments in Germany have created special programs to support the implementation of Industrie 4.0. And these political actors give sense in the form of explanations and sensemaking opportunities for others. For example, the German Federal Ministry of Education and Research (BMBF) states:

"The future project of Industrie 4.0 aims at enabling the German industry to be prepared for the future of production." (BMBF website[15])

And the German Federal Ministry for Economic Affairs and Energy (BMWi) writes:

"The German Government seeks to actively promote and shape the transition into the digital era. 'Industry 4.0'—networked production—has the potential to significantly influence Germany's key industrial sectors. It is vital to support small and medium-sized enterprises in enhancing their innovative capacity through new digital technologies." (BMWi website)[16]

Large companies followed suit. Companies like SAP, Siemens and Bosch were among the first to become actively involved in the debate on Industrie 4.0. They used Industrie 4.0 as part of their strategic activities and to create some urgency for their own business interests. Small and medium-sized companies followed later and more hesitantly—at least in the beginning. Companies have been the main target for the sensegiving activities of industry associations and political actors. The reasons why companies join their efforts vary: Several different incentives have been created for companies, and industry associations and political actors have encouraged adoption. The media have also reported on Industrie 4.0,

---

[15]https://www.bmbf.de/de/zukunftsprojekt-industrie-4-0-848.html, accessed June 22, 2016.
[16]http://www.bmwi.de/DE/Themen/Industrie/industrie-4-0.html, accessed 08.08.2016.

as though it were a certifiable fact, and not a vision. Many businesses believe that they need to innovate to stay competitive, but often they are not sure how. "Industrie 4.0" lays out a pathway to continued competitiveness *and* a recipe for innovation. For many companies, Industrie 4.0 has served as a guide to prepare for the future and demonstrate to shareholders, partners, banks, and so on, that management is on top of the latest developments.

The core of the emerging field around Industrie 4.0 was formed by industry associations, political actors, and companies. Less central, but still part of the field, are the media, universities, and labor unions. The media have contributed to the field by distributing the general idea of Industrie 4.0 and making more and more people aware of this envisioned future. Others situated outside the core of the organizational field could then become aware of the debate. Universities contributed to the discussion by framing their research in terms of Industrie 4.0, thus contributing to the narrative, but also making engineering students aware of its existence. Virtually all newly trained engineers now "know" about Industry 4.0 and its importance. Professional training is therefore another important diffusion mechanism (Meyer 2013). Finally, labor unions use this vision as a reference and reason to discuss the future of work and labor.

By addressing and involving these groups, Industrie 4.0 integrated different areas of society which require an above-average level of explicit sensemaking and sensegiving activities. Managers and politicians more or less embody the ideal type of actors who carry out specific scripts that rely heavily on sensemaking and sensegiving elements. They are expected to act strategically, find innovative solutions to existing and future problems, and, in the process, secure future success. All the while, these actors must explicate and justify their sensemaking activities. Furthermore, they are expected to embrace science and technology in all of these endeavors. Active references to Industrie 4.0 by very different actors early its career enabled the widespread diffusion of the term.

At this point, it is important to reiterate that the diffusion of Industrie 4.0 as an envisioned future did not go hand in hand with a homogenization of the different actors' views. The diffusion of the term and its use by many actors did not result in a convergence of the interpretations of this envisioned future. Like Hoffman (1999) emphasizes for organizational fields: Different actors share an issue, but their different interpretations of that issue continue to exist. This multiplicity is supported by the properties of envisioned futures outlined above as well as the narrative of Industrie 4.0. While quite specific in some regards (e.g. in linking industry to cyber-physical systems), Industrie 4.0 still manages to remain quite open and vague. It allows for multiple interpretations and framings. Actors utilize this potential as they frame Industrie 4.0 based on their own relevance structures and sensemaking processes. In our interviews we asked respondents what Industrie 4.0 meant to them and the organization they represented. Some quotes from our interviews illustrate the range of responses as well as the very different levels addressed by their interpretations.

One company representative, for example, considered Industrie 4.0 to be relevant within the company as it describes a new stage of optimization:

"Industrie 4.0 describes an overarching optimization. Before, you had individual systems, then you optimized more than one system and now you optimize all of them at once. And that is the essential aspect that sets it apart from earlier discussions." (interview, company representative)

An industry association representative described Industrie 4.0 as a change in customer relations:

"What's special about Industrie 4.0 is that it allows industrial companies to be in constant contact with their customers." (interview, industry association representative)

Another interviewee saw it as a process optimization as well, but also considered it relevant on a broader, in this case, national level:

"For us, it means process optimization in order to keep German SMEs competitive." (interview, company representative)

Yet another level is addressed by a company that considers Industrie 4.0 the real-time simulation of material production processes (digital twin):

"Industrie 4.0 is the virtual mapping of all instances and all steps of the production process. Each element of production, every machine, has a digital twin." (interview, company representative)

Other respondents considered the impact of Industrie 4.0 on people and the way they work:

"Industrie 4.0 allows people – for the very first time – to truly be independent of time and space in their work." (interview, company representative)

Additional respondents described the envisioned future as "batch size 1," "a new connection on a large scale," "capacity management between different regional units or different companies," "feedback for all our own products," "[the ability to] combine my product with intelligent services," or "the connection between different elements of production, that is, machines, products and the workpiece." Others considered it a marketing strategy, a new framing of developments that were already underway, or a description of something that has not (yet) arrived:

"Industrie 4.0 is the same product in a different wrapping. No, not even wrapping: different advertising." (interview, company representative)
"From our perspective, the actual development of Industrie 4.0 is much slower than what you hear in the media and at conferences. For now, it's primarily a search effort." (interview, labor union representative)
"My impression is that Industrie 4.0 is a kind of virtual term that needs to be filled with substance. It's primarily a projection. People are basically using it to say, 'Hey, some big changes are happening. And we need to get on board in our industry. We cannot miss this.' Maybe it's also driven by the fear that the big internet companies might rule everything someday." (interview, workers' council representative)

Some people we interviewed saw the biggest impact of Industrie 4.0 in the effect it had on other areas, for example on the future of work:

"The primary impact we see it that it has revitalized an active labor market policy." (interview, workers' council representative)

In one interview, a representative of a large industry association described how he and his colleagues had tried to pin down what Industrie 4.0 really means:

"So, there is the question: What is Industrie 4.0? And [our industry association] took a survey of its different uses. We stopped at 134 or 135. We stopped counting. There are just so many definitions out there." (interview, industry association representative)

Therefore, with the ongoing diffusion of this envisioned future, the number of corresponding interpretations has only increased. What is striking is that these differences extend well beyond the technological foundation of Industrie 4.0. Moreover, the above interpretations refer to very different levels, from aspects of the production process, to intra-organizational improvements, to inter-organizational ties and even changes on a national or global scale. The technology itself is at turns specific, and at turns vague. Some consider Industrie 4.0 to be a marketing term—mere window dressing—while for others it is quite technical. In line with these findings, Acatech, and consulting agencies, have developed a variety of checklists to evaluate the implementation level of Industrie 4.0 that an individual company has already achieved. Because this envisioned future can mean so many things and so many different levels, the task of evaluating whether Industrie 4.0 itself is realistic or credible is extremely complicated. As the primary feature of an envisioned future, this versatility compounds the problem. The common ground in this case is the story and meaning of Industrie 4.0 and not the specific uses of associated technology or even its level of relevance.

Field formation, or the "structuration" of a field as DiMaggio and Powell (1983) put it, goes beyond the development of links between existing organizations. Within the field, an envisioned future creates orientation, values, and new forms of organization. This can also be seen in the case of Industrie 4.0. Some actors within the field—those with a technological understanding of Industrie 4.0—have developed technical standards and indicators for Industrie 4.0 and its implementation. For example, associations such as TÜV Süd or the Plattform Industrie 4.0 and acatech have already produced such content. Certain funding schemes which address Industrie 4.0 and connections between various actors to achieve this goal now require applicants to explicitly account for such standards.[17]

Another indicator of the stabilization of the envisioned future and its associated field is the creation and formation of specific (meta)organizations to support Industrie 4.0 ideas. By now, a variety of organizations have been created with the explicit purpose of advancing the Industrie 4.0 transformation. A good example of this is the Plattform Industrie 4.0, which justifies its existence by referencing the envisioned future:

"How can Germany become the leading factory equipment supplier for Industrie 4.0? How can Germany further improve its competitiveness as a production location through Industrie 4.0? What role can Germany play in setting standards and how can Industrie 4.0 benefit people in the world of work? Plattform Industrie 4.0 aims to find answers to these questions through dialogue." (Plattform Industrie 4.0)

Once special organizations are in place whose existence depends on and is dedicated to the success of an envisioned future, there are also very committed actors who strive to see it through. Because, in short, their organization's success—or perhaps its very existence—is tied up in the success of the envisioned future of Industrie 4.0. Members of these organizations work hard to make Industrie 4.0 a reality—not only for the sake of the vision, but also in the interest of their jobs and careers.

---

[17]An example is a funding scheme bei the German Ministry of Education and Research: https://www.bmbf.de/foerderungen/bekanntmachung-1352.html, accessed November 21, 2017.

Existing organizations restructure themselves to adapt to the envisioned future. This can lead to the creation of new departments.

"Right now, we are in the midst of creating a new department. There will be around 50 people working there on the topic of Industrie 4.0. They will all be in the new building, next to the management. So we [the board members] will be right in the middle of it to make sure we do not miss anything." (interview, company representative)

These two elements—new organizations and restructuring of existing ones—can go hand in hand. For example, most industry associations have created internal structures dedicated to the promotion of this specific envisioned future, which—moreover—follows the structure of the newly created Plattform Industrie 4.0:

"So, for every working group in Plattform Industrie 4.0, we created a mirror body within our association. And the other industry associations did the same." (interview, industry association representative)

But the platform itself has also been massively restructured. When the dynamics of the envisioned future changed, the platform's internal structure, including the structure of its members, was reorganized to account for these changes.

Another indicator for the staggering success of Industrie 4.0—and for the limits of its control—is that even the people who actually participated in its creation and early diffusion are surprised by the current state of affairs:

"If someone would have asked me in early 2012, I wouldn't have expected this. I would have said that it's interesting, but that's about it. And today I am getting invited to China and I travel to Beijing or Shanghai or Tokyo or New York and when I say "Industrie 4.0," everybody knows what I'm talking about – I never would have expected this back in 2012". (interview, industry association representative)

These dynamics have led to the stabilization of an issue field for Industrie. 4.0. From something that needed careful and repeated explaining and framing, the future portrayed by Industrie 4.0 has now transformed a rather factual description of the developments required of both companies and political actors. Organizations are now called on to justify their point of view if they do *not* want to get behind this vision of the future, and not the other way around. The envisioned future has influenced a variety of organizations and is reflected in their organizational structures.

The envisioned future of "Industrie 4.0" is very well established in Germany. Its institutionalization has progressed to the point that it is often cited as a fact, not an option or an idea. Industrie 4.0 no longer requires any justification; instead, it can be used to justify other phenomena. Companies themselves now use "Industrie 4.0" in their sensegiving activities to gain legitimacy and make others understand and accept their activities:

> "We advertise things differently now. We highlight the keyword "Industrie 4.0". About three years ago, we exhibited our technology at a fair and nobody cared. Now, the product is the same, but we changed the packaging. Now, we call it "Industrie 4.0" and the people are lining up to get it." (interview, company executive)

When framed as part of Industrie 4.0, the activities and products of this company became more interesting and more relevant to others. In other words, a field has formed and with it a shared understanding of the relevance of Industrie 4.0.

The number 4.0, similar to versioned software products, has become a widespread signifier of innovation and progress, of being "up to date". The use of 4.0 has also extended far beyond the context of industrial change and can be found in a variety of different contexts. There is a discussion on *work 4.0*, as well as *economy 4.0* and *capitalism 4.0*. There is *consulting 4.0*, a book on *Deutschland 4.0* and a political campaign for a *Nordrhein-Westphalen 4.0*. There is an *Apotheke 4.0* and a book on *water 4.0*. We also have *space 4.0*, *sport 4.0*, and *terrorism 4.0*. And, of course, if anyone talks about the future, it is likely to be *future 4.0*.

8   Concluding Remarks

The model of envisioned futures presented in this article offers an analytical perspective on the early phases of envisioned futures and how they influence the present. It highlights the relevance of organizations and organizational fields in such processes. It suggests a dynamic model for the analysis of the creation, emergence and diffusion of envisioned technological futures. This model draws on and combines concepts from STS such as imaginaries and expectations in technological development with concepts from institutional and organizational theory such as a field-based perspective and Weick's elaborations on sensemaking in organizations. Based on these theoretical concepts, the development of envisioned futures is characterized as a series of iterations involving organizational sensemaking, enactment, sensegiving, institutionalization, and field formation. The proposed model shows how an envisioned future can flourish when it is picked up and used by different organizations and it acquires meaning beyond individual organizations. Such a technology-based vision of the future can only be successful if it proves useful to a variety of different organizations that represent different rationalities in society. It must provide support to all of these organizations in their sensemaking and sensegiving activities. The envisioned future must prove itself as a tool to justify and legitimize activities. Pivotal to this model is the argument that envisioned futures, like Industrie 4.0, are very useful for organizations, because they help them to reduce uncertainty and enable action. Envisioned futures do this not by providing additional information, but by offering an interpretative framework for information that is already "out there" in the world. At the core of such an envisioned future are not the quality, performance or "factual" potential of specific technologies. Though technologies are highlighted in the narrative, what cinches that narrative's success its adaptiveness or its ability to remain vague and therefore useful to very different actors. Though their individual sensemaking activities, actors can thus reduce uncertainty and take action. Envisioned futures present a sensemaking mechanism that is established on the level of an organizational field.

This article has also shown the limits of envisioned futures. For example, they are not a very precise tool for strategic action. Even the actors creating an envisioned future cannot control its development to a degree that would enable intentional steering. If and when an organizational field forms, its basic issue can be framed and interpreted in very different ways. As with path dependent processes in general (Meyer 2013), if an envisioned future is successful, so many actors will have contributed to its development and interpretation that it is virtually impossible to take control of that future—even when a large group of actors attempts to take the reins. The envisioned future of Industrie 4.0, for example, triggered a debate on the future of work (often coined *Arbeit 4.0*, i.e. work 4.0), which frequently encroaches on the envisioned future of Industrie 4.0. The creators of the original envisioned future did not intend for these and other interventions to emerge, nor can they control them.

The model suggested in this paper is intended as a contribution to the models of socio-technical futures in STS (science and technology studies). It provides an analytical framework which accounts for the role of organizations in the processes that shape envisioned futures and their diffusion and realization. The model focuses on the early phases of such futures and suggests criteria to discern how envisioned futures become successful. It shows the dynamics of such futures before they become "shared by many"—which has been the dominant reference point so far. This paper also makes contributions to organization studies. First, it shows how newly founded organizations can play a critical role in processes of institutionalization. Second, it shows, how a specific kind of field aimed at supporting new forms of science and technology emerges. In addition, it offers a starting point for an analysis of the relationship between sensemaking, institution-alization, and organizational fields. While emphasizing individual organizations in the process of sensemaking, such a perspective is not limited to this level. An envisioned future stabilizes and diffuses as a result of different sensemaking processes and their dynamics. Such processes take place on a level beyond individual organizations and beyond the control of individual organizations. New connections between actors develop, shared understandings spread, and a field emerges.

With few exceptions, findings in organization studies have been ignored by STS scholars—and vice versa. This paper wants to help create a more durable bridge between these two fields, which would profit immensely from a more frequent and in-depth exchange.

References

Ahrne, G., & Brunsson, N. (2005). Organizations and meta-organizations. *Scandinavian Journal of Management, 21*(4), 429–449.

Arthur, W. B. (1988). Self-reinforcing mechanisms in economics. In W. B. Arthur (Ed.), *The economy as an evolving complex system* (pp. 9–31). Boston: Addison-Wesley.

Arthur, W. B. (1989). Competing technologies, increasing returns, and lock-in by historical events. *The Economic Journal, 99,* 116–131.

Bakker, S., von Lente, H., & Meeus, M. T. H. (2012). Credible expectations. The US department of energy's hydrogen program as enactor and selector of hydrogen technologies. *Technological Forecasting & Social Change, 79*(6), 1059–1071.

Balogun, J., & Johnson, G. (2004). Organizational restructuring and middle manager sense-making. *Academy of Management Journal, 47*(4), 523–549.

Berger, P. L., & Luckmann, T. (1967). *The social construction of reality: A treatise in the sociology of knowledge.* London: Penguin.

Besio, C., & Meyer, U. (2015). Heterogeneity in world society. How organizations handle contradicting logics. In B. Holzer, F. Kastner, & T. Werron (Eds.), *Isomorphism and differentiation: From globalization(s) to world society* (pp. 237–257). New York: Routledge.

Boje, D. M. (1991). The storytelling organization: A study of story performance in an office-supply firm. *Administrative Science Quarterly, 36*(1), 106–126.

Brown, A. D., Stacey, P., & Nandhakumar, J. (2008). Making sense of sensemaking narratives. *Human Relations, 61*(8), 1035–1062.

Currie, G., & Brown, A. D. (2003). A narratological approach to understanding processes of organizing in a UK hospital. *Human Relations, 56*(5), 563–586.

Dierkes, M. (1988). Organisationskultur und Leitbilder als Einflußfaktoren der Technikgenese: Thesen zur Strukturierung eines Forschungsfeldes. In ISS Forschung (Ed.), *Ansätze sozialwissenschaftlicher Analyse von Technikgenese* (pp. 49–62). München: Institut für sozialwissenschaftliche Forschung (ISF).

Dierkes, M., Hoffmann, U., & Marz, L. (1996). *Visions of technology*. New York: Campus.

DiMaggio, P., & Powell, W. W. (1983). The iron cage revisited: Institutional isomorphism and collective rationality in organizational fields. *American Sociological Review, 48,* 147–160.

Dunford, R., & Jones, D. (2000). Narrative in strategic change. *Human Relations, 53*(9), 1207–1226.

Fligstein, N., & McAdam, D. (2012). *A theory of fields*. New York: Oxford University Press.

Fujimura, J. (1988). The molecular biological bandwagon in cancer research: Where social worlds meet. *Social Problems, 35,* 261–285.

Gabriel, Y. (2004). Narratives, stories and texts. In D. Grant, C. Hardy, C. Oswick, & L. Putnam (Eds.), *The Sage handbook of organizational discourse* (pp. 62–77). Thousand Oaks: Sage.

Gephart, R. P. (1993). The textual approach: Risk and blame in disaster sensemaking. *Academy of Management Journal, 36*(6), 1465–1514.

Gioia, D., & Chittipeddi, K. (1991). Sensemaking and sensegiving in strategic change initiation. *Strategic Management Journal, 12*(6), 433–448.

Gladwell, M. (2000). *The tipping point: How little things can make a big difference* (1st ed.). Boston: Little Brown.

Granovetter, M. (1978). Threshold models of collective behavior. *American Journal of Sociology, 83*(6), 1420–1443.

Hirsch-Kreinsen, H. (2016). Arbeit und Technik bei Industrie 4.0. *Aus Politik und Zeitgeschichte, 66*(18–19), 10–17.

Hoffman, A. J. (1999). Institutional evolution and change: Environmentalism and the U.S. chemical industry. *Academy of Management Journal, 42*(4), 351–371.

Humphreys, M., & Brown, A. D. (2002). Narratives of organizational identity and identification: A case study of hegemony and resistance. *Organization Studies, 23*(3), 421–447.

Isabella, L. A. (1990). Evolving interpretations as a change unfolds—How managers construe key organizational events. *Academy of Management Journal, 33*(1), 7–41.

Jasanoff, S., & Kim, S.-H. (2009). Containing the atom: Sociotechnical imaginaries and nuclear power in the United States and South Korea. *Minerva, 47,* 119–146.

Kagermann, H., Lukas, W.-D., & Wahlster, W. (2011). Industrie 4.0: Mit dem Internet der Dinge auf dem Weg zur 4. industriellen Revolution. *VDI Nachrichten, 13,* 2.

Lorenz, P. (2017). *Digitalisierung im deutschen Arbeitsmarkt. Eine Debattenübersicht*. Sankt Augustin: Konrad-Adenauer-Stiftung e. V. und Stiftung Neue Verantwortung e. V.

Maitlis, S., & Christianson, M. (2014). Sensemaking in organizations: Taking stock and moving forward. *The Academy of Management Annals, 8*(1), 57–125.

Meyer, U. (2013). Self-reinforcing mechanisms in organizational fields: The development of an innovation path in the car industry. In J. Sydow & G. Schreyögg (Eds.), *Self-reinforcing processes in and among organizations* (pp. 17–34). Basingstoke: Palgrave Macmillan.

Meyer, U. (2016). *Innovationspfade. Evolution und Institutionalisierung komplexer Technologie*. Wiesbaden: Springer VS.

Pfeiffer, S. (2015). Warum reden wir eigentlich über Industrie 4.0? Auf dem Weg zum digitalen Despotismus. *Mittelweg 36, 24*(6), 14–36.

Sackman, S. A. (1991). *Cultural knowledge in organizations: Exploring the collective mind*. Newbury Park: Sage.

Sandelands, L. E., & Stablein, R. E. (1987). The concept of organizational mind. In S. Bacharach & N. DiTimaso (Eds.), *Research in the sociology of organizations 5* (pp. 135–161). Greenwich: JAI Press.

Starbuck, W. H., & Milliken, F. J. (1988). Executives' perceptual filters: What they notice and how they make sense. In D. C. Hambrick (Ed.), *The executive effect: Concepts and methods for studying top managers* (pp. 35–65). JAI Press: Greenwich.

Swanson, E. B., & Ramiller, N. C. (1997). The organizing vision in information systems innovation. *Organization Science, 8*(5), 458–474.

Taylor, J. R., & Van Every, E. J. (2000). *The emergent organization: Communication as its site and surface*. Mahwah: Erlbaum.

van Lente, H. (2000). Forceful futures: From promise to requirement. In N. Brown, B. Rappert, & A. Webster (Eds.), *Contested futures: A sociology of prospective techno-science* (pp. 43–64). Aldershot: Atheaneum.

van Lente, H., & Rip, A. (1998a). Expectations in technological developments: An example of perspective structures to be filled in by agency. In C. Disco & B. van der Meulen (Eds.), *Getting new technologies together* (pp. 203–229). New York: De Gruyter.

van Lente, H., & Rip, A. (1998b). The rise of membrane technology: From rhetorics to social reality. *Social Studies of Science, 28*(2), 221–254.

Weick, K. E. (1995). *Sensemaking in organizations*. Thousand Oaks: Sage.

Weick, K. E., & Roberts, K. H. (1993). Collective mind in organizations: Heedful interrelating on flight decks. *Admin-istrative Science Quarterly, 38*, 357–381.

Weick, K. E., Sutcliffe, K. M., & Obstfeld, D. (2005). Organizing and the process of sensemaking. *Organization Science, 16*(4), 409–421.

Windeler, A., Knoblauch, H., Löw, M., & Meyer, U. (2018). Innovationsgesellschaft und Innovationsfelder. In J. Hergesell, C. Minnetian, A. Maibaum, & A. Sept (Eds.), *Innovationsphänomene. Modi und Effekte der Innovationsgesellschaft* (pp. 17–38). Wiesbaden: Springer VS.

Uli Meyer (Ph.D.) is sociologist, and head of the research group "reorganizing industries" at the Munich Center for Technology in Society (MCTS) at Technische Universität München and acting professor (Vertretungsprofessur) for sociology of digital work at the Ruhr Universität Bochum.

Between Past, Present, and Future— The Temporality of Sociotechnical Futures in India's GM Crops Debate

Andreas Mitzschke

Abstract

This chapter explores the enduring controversy about genetically modified crops in India. It asks what role different interpretations of past agricultural development play in the construction of contested sociotechnical futures with(out) transgenic crops. The theoretical frame for this analysis combines the 'social construction of technology' with the concept of 'sociotechnical imaginaries' to understand temporal scales in the construction of such futures. Based on document analysis and semi-structured interviews, the chapter identifies four sociotechnical imaginaries in the debate. These are shaped by different interpretations of India's agricultural past. The author argues for a greater role of temporality in studying technological controversies.

A. Mitzschke (✉)
Faculty of Arts and Social Sciences, Maastricht University,
Maastricht, The Netherlands
e-mail: andreas.mitzschke@maastrichtuniversity.nl

© Springer Fachmedien Wiesbaden GmbH, part of Springer Nature 2019
A. Lösch et al. (eds.), *Socio-Technical Futures Shaping the Present*,
Technikzukünfte, Wissenschaft und Gesellschaft / Futures of Technology,
Science and Society, https://doi.org/10.1007/978-3-658-27155-8_7

1   Introduction

This chapter explores the role of constructed futures in the public debate about genetically modified (GM) crops in India.[1] Ideas about the impact of technological change on the future do not only reflect the fears and hopes of various actors about how society might evolve, but they also reflect their concerns with the contemporary social and political state of society. The enduring controversy about GM crops in India is an exemplary case of a technological controversy in which various actors construct competing, conflicting, and contested futures with the help of science and technology. These futuristic visions are usually understood to revolve around the potential risks and benefits of transgenic crops. In the following, I want to explore the construction of such futures in the Indian debate about GM crops with the help of the Science and Technology Studies (STS) concept of sociotechnical imaginaries. Sociotechnical imaginaries are collectively held and publicly performed visions of desirable futures, rooted in shared understandings of social life and order made possible by science and technology (Jasanoff and Kim 2009, 2013, 2015).

Such shared understandings of sociotechnical reality and constructions of futuristic visions connect the present to the future. The central questions I aim to address are: How can we conceptualise the connection between constructed technological futures and present social realities in India's GM crops debate? What can a temporal perspective of this relationship add to studying debates about technological risks? Based on a textual analysis of documents and semi-structured interviews with key actors of the Indian debate about transgenic crops, I describe what sort of sociotechnical imaginaries the actors construct, and how these connect various time scales.[2] I argue that the proponents and opponents

---

[1]Genetic modification of plants is more precisely described as transgenic modification, which refers to recombinant DNA technique, i.e. inserting genes extracted from one species (e.g. a bacterium) into the DNA of a target organism (e.g. a plant) when both organisms are sexually incompatible. This intervention is done using complex laboratory techniques that allow manipulating organisms on the molecular level. Accordingly, I will use the terms 'genetically modified' and 'transgenic' interchangeably in this chapter.

[2]I draw on material I collected during two ethnographic visits to India from February to April 2012, and January to March 2013. My field visits were funded by the Faculty of Arts and Social Sciences, Maastricht University as part of my dissertation research which focuses on the role of publics in the Indian and European controversies about GM crops and the co-shaping of technological development and democratic politics.

of transgenic seeds situate the *futures* they imagine with or without the technology within their interpretation of *past* agricultural development. Accordingly, constructing futures is as much related to interpretations about the past, as to the shaping of the present; i.e. the concept of sociotechnical imaginaries is useful for understanding the temporal relations of the GM crops debate between past, present, and future. Such a temporal perspective widens the analysis of the debate about GM crops beyond the frame of 'technological risk': the normativity actors connect to transgenic seeds, the political role of the agrarian community in shaping technological development, and discourses about modernisation.

My theoretical point of departure is the social constructivist perspective of science and technology, which regards scientific knowledge as well as technological artefacts as socially constructed (Bijker 1995; Collins 1983; Pinch and Bijker 1984; Shapin and Schaffer 1985). In particular, I look at technology through the lens of the social construction of technology (SCOT) (Pinch and Bijker 1987) perspective, which explains how the same artefact can mean different things to different groups of people. SCOT endeavours to sociologically deconstruct these meanings. So-called 'relevant social groups' share the same set of interpretations they ascribe to a technology, such as problem definitions and suggested solutions. The interaction between different relevant social groups, in particular in technological controversies, reveals the "interpretative flexibility of technology" (Pinch and Bijker 1987), i.e. technologies are subject to a culturally framed process of construction and interpretation. There is more than one interpretation of what a technology is or does, and whether it works or not; the same technology has different meanings to different relevant social groups. SCOT aims to explain how these meanings differ, and under which circumstances such meanings converge (or not), eventually leading to the closure (or continuation) of technological controversy.

I complement this perspective with the concept of sociotechnical imaginaries (hereafter STIs) which explicitly focuses on the imaginative repertoires that actors come up with in constructing futures with science and technology. The concept was initially developed in anthropology (Marcus 1995), but has recently been picked up in STS to explain "how different imaginations of social life and order are co-produced along with the goals, priorities, benefits, and risks of science and technology" (Jasanoff and Kim 2009, p. 141). STIs are simultaneously about how actors construct techno-scientific futures and about their ideas how the social world should be ordered.

In accordance with social constructivist perspectives on science and technology, sociotechnical imaginaries are a way of looking at the relationship between science, technology and society: science and technology do not shape the social unidirectional, but technology is socially shaped. There is a co-construction between knowing about and living in the world: "knowledge and its material embodiments are at once products of social work and constitutive of forms of social life" (Jasanoff 2004, pp. 2–3). So, as ideas about the future shape and get shaped by technological development, we can equally ask how the sociotechnical imaginaries of GM crops shape and get shaped by the actors' present way of life. My argument is that looking at the GM crops debate through the lens of STIs provides us with a fruitful perspective on the connections between the past, the present, and constructed futures. This means we have to look at the entire temporal scale actors invoke in constructing futures with science and technology in order to understand how they work to shape the present. The temporal perspective STIs enable, accordingly allows to look at the debate about the risks of transgenic crops from a different angle to pay more attention to the wider issues the actors invoke in the Indian controversy.

Let me commence with describing how sociotechnical imaginaries serve to understand the sociotechnical futures which relevant social groups construct in the current debate about GM crop technology in India. I first depict the background to the Indian controversy. Then I describe four different STIs that revolve around the risks and benefits of GM crops, and I explain how these ideas about the future are rooted in the social groups' interpretations of the past. In the final section I discuss how a temporal perspective point to the limits of risk discourses for understanding the Indian debate. A temporal perspective instead highlights how the debate entails wider contestations about normativity, community, and modernity. I will conclude by pointing to how the temporality of sociotechnical imaginaries entails theoretical implications for STS research.

2   Competing, Conflicting, Contested Futures in India's GM Crops Debate

In 2002, the introduction of insect-resistant Bt-cotton as the first commercial GM crop in India led to intense public controversy about the technology's diffusion, performance, and its social and environmental appropriateness (Shah 2005). The attempted marketization of India's first GM food crop, insect-resistant Bt-brinjal (eggplant), was met with even fiercer public opposition and debate.[3] A nation-wide

---

[3]Insect-resistance traits are induced into plants with transgenes from the soil bacterium bacillus thuringiensis (Bt). If successfully modified, the plant synthesizes a protein that works as a pesticide against lepidopteran insects, such as the cotton bollworm (Bt-cotton) or the Brinjal Fruit and Shoot borer (Bt-brinjal).

public consultation process induced by the Minister of Environment and Forests resulted in the imposition of a moratorium on Bt-brinjal in 2010. Since each side of the Indian controversy engages in its own "authentication loops" (Stone 2012) for generating, validating, and publicising research data on the agronomic performance of transgenic crops, useful comparisons are difficult to make. Hence, the debate about transgenic crops is rife with speculative ideas about the future with or without this technology. Such futures are commonly understood to revolve around the alleged risks and benefits of GM crops: to human health and the ecosystem, to the stability of the agrarian economy, and the potential social and economic consequences of introducing the technology to the Indian farming system.

As the common sense notion of imaginaries suggests, sociotechnical imaginaries are ways actors conceive of a future shaped by science and technology. The concept has been used to describe state institutions' and actors' mechanisms of governance, similar to agenda setting in conventional policy analysis (Hajer 1995; Smith 2009). Although the concept was initially developed to describe how collectively imagined forms of social life and scientific and technological development are intertwined with national political cultures (Jasanoff and Kim 2009, 2013), other actors than nation states, such as institutions, companies, environmental NGOs, and expert bodies construct their own imaginaries (Jasanoff and Kim 2015). Moreover, imaginaries about science and technology are closely intertwined with prescriptive notions of the good of the social world and the political community: "techno-scientific imaginaries are simultaneously also 'social imaginaries', encoding collective visions of the good society" (Jasanoff and Kim 2009, p. 123), hence: sociotechnical imaginaries.

If different social groups attach different meanings to the same technology (*interpretative flexibility of technology*), then their ideas about society's techno-scientific future should be as diverse. Understanding techno-scientific controversies from the perspective of imaginaries is fruitful because it allows scrutiny of the performative and representational dimensions of such debates. In my analysis, I locate sociotechnical imaginaries as futuristic visions according to (1) their representation of risks and benefits, (2) their instrumental character in calling for targets of public action (i.e. policy preferences), and (3) their relation to symbolic articulation of interpretations of the past. Doing so, I will describe how the relevant social groups of the GM crops debate in India construct a multiplicity of competing, conflicting, and contested imaginaries.

Jasanoff and Kim suggest that little attention has been paid so far to studying how sociotechnical imaginaries make and sustain the focus on some risks and benefits of sociotechnical change, while systematically downplaying other ideas about collective risk taking (2013, p. 195). My take on the temporality of sociotechnical imaginaries, i.e. how various ideas about future sociotechnical development rest on diverse interpretations of the past, aims to draw attention to those issues that a risk-based discourse marginalises or silences. To understand the relationship between sociotechnical imaginaries and the discourse on risk, we must not only look at policy documents and narratives, but we also need to include the actor's symbolic and cultural resources with which they frame and represent alternative futures, such as images, texts, memories, and metaphors (Jasanoff and Kim 2015, p. 37). To study the symbolic and representational character of imaginaries, I draw not only on documents, but also on semi-structured qualitative interviews I conducted with activists and campaigners, civil society organisations, industry advocates, NGO representatives, and scientists during two field visits to India in the years 2012 and 2013. Semi-structured interviews are a good method to understand actors' world views, assumptions, and meaning constructions (Roulston 2010).

The four sociotechnical imaginaries I identified in the construction of futures in the GM crops debate are: productivity, environmentalism, complex ecology, and seed sovereignty. Based on my analysis of empirical data, the following sections describe how the actors of the Indian debate represent risks and benefits of transgenic crops, what calls on public policy they imply, and how they articulate interpretations of past agricultural development. In doing so, they construct various sociotechnical futures, which in turn reflect on their experience of present social reality.

2.1 Productivity

The proponents of GM crops present transgenic crops as a solution to one of the most pressing problems they assign to agriculture: productivity. The construction of a future in which transgenic crops are the panacea to solve the problem of hunger and malnutrition by boosting agricultural productivity makes the sociotechnical imaginary of productivity. For instance, Monsanto company alludes that with a growing world population the task for agriculture is to produce higher quantities of food crops on less arable land, an "objective that cannot be realised without GM crop technology" (Grant 2008; cf. Monsanto 2012). In India, food security is a pressing problem that is rooted in the historical experience of famine and hunger which in turn activates collective consciousness and motivates policy targets (Pritchard et al. 2013). Political leaders such as president Pranab Mukherjee (Pritchard et al. 2013), Union Minister Sharad Pawar (Express 2013), industry advocates (Rao 2013), and biotechnologists (Padmanaban 2013) frequently invoke a future in which transgenic crop technology can solve the problem of food security by increasing agricultural productivity. For instance, Professor Padmanaban, a molecular biologist at the Indian Institute of Science in Bangalore, who contributes frequently to the GM crops debate in public media, emphasised in an interview: "the fact remains that we need to produce more food in accordance with the rate of population growth". He sees GM crops as able to provide a range of crop traits useful for productivity in the future, such as drought tolerance and enhanced nutritional composition: "More production and better nutrition, would that not make for a revolution in a country like India?" (interview Padmanaban 2012).

The sociotechnical imaginary of productivity also constructs a future in which transgenic crops enable Indian farmers to produce for the global market and to boost India's economic competitiveness: "increase in productivity is in the interest of all, it is a shared goal" (industry representative at 'Doubling Food Production in India in Five Years', Conference, Delhi, 2013). Transgenic crops, therefore, become part of a future in which India harmonises with global practices of technology-based innovation and competitiveness. The STI of productivity therefore is twofold: GM crops are envisioned to solve the problem of food shortages, and to enable Indian farmers to contribute to overall economic growth by tapping into global agricultural commodity markets. The policy imperative implied by the imaginary of productivity is to put more trust into science and technology developers and to harmonise and ease the demands of the regulatory system for product safety and risk assessment (interview Rao 2013).

With such benefits in mind, proponents of GM crop technology regard the potential risks as negligible and the regulatory mechanisms for product safety as too strict: "What are those risks? [...] Any of them has no meaning at all, absolutely there is no meaning. Because look at malnutrition, look at starvation, all happening in this country" (interview Padmanaban 2012). The imperative of the altruistic notion of feeding the hungry by means of better technology accordingly supersedes any concern for food safety. If risks still become a problem, Padmanaban alludes, "we come up with new things [...] we can manage it like with other problems". The proponents of transgenic crops regard it as a precise technological intervention that can cope with risks through further technological adaptation and refinement. Yet, the proponents of GM crops do not only construct a future where transgenic seed technology contributes to higher productivity, but also to making agriculture more environmentally friendly. Before I explain how the STI of productivity is rooted in a specific interpretation of India's agricultural history, let me explain the construction of what I call the sociotechnical imaginary of environmentalism.

2.2   Environmentalism

The proponents of transgenic crops invoke it as part of a future with a decreased environmental impact of agriculture. Scientist Padmanaban contends that insect-resistant GM crops yield environmental benefits: "Bt-crops are the best option for organic farming. The function of Bt-crops is to reduce pesticide sprays (interview Padmanaban 2012). An Indian industry representative from the Association of Biotech-Led Enterprises (ABLE) agrees: "because there are fewer pesticide sprays, environmentally it is safer" (interview Seetharama 2013). Also, these actors contend that realising such benefits will lead to a greater public acceptance of GM crop technology. They support their claims by referencing field-level data that show an overall reduction in pesticide application and its associated positive environmental impacts (e.g. Choudhary and Gaur 2012). The prospects for a sociotechnical future in which pests are easily controlled by transgenic crops, and without external pesticide applications, is supported by studies provided by advocacy organisations such as the International Service for the Acquisition of Agri-Biotech Applications (ISAAA) which states the potential of reducing pesticide applications in Bt-cotton up to 40 per cent (Choudhary and Gaur 2012). Other studies produced by scientists with links to the biotech-industry also support the view that Bt-crops will positively contribute to environmental sustainability (Carpenter 2011).[4] Yet such studies remain remarkably silent on the scientific uncertainties of understanding the complex relationships between various kinds of insects in agricultural fields, and the impact Bt-crops might have on them, or the possible environmental consequences of the transgenes crossing with wild relatives or non-GM crops (e.g. Hammond 2010). Yet, the policy implications implied by such ideas about the many benefits GM crops might have for the environment are clear: to have less stringent environmental risk assessment because the proponents regard the existing risk assessment regime as having provided sufficient evidence for the environmental safety of transgenic crops.

How are the sociotechnical imaginaries of productivity and environmentalism rooted in interpretations of the past? Actors constructing such futures in which GM crops increase agricultural productivity, while at the same time reducing the environmental impact of pesticide applications, unanimously refer to the Green

---

[4]Jane Carpenter works for Croplife International a research organisation that is financed by and can be said to represent the interests of biotechnology companies in the transgenic seed sector such as BASF, Bayer CropScience, Dow AgroScience, Monsanto, and Syngenta (CropLifeInternational 2014; see also Powerbase 2014).

Revolution period as a successful technological project.[5] The Green Revolution refers to the period from the early 1960s in which improved agricultural technologies such as high yielding crop varieties, inorganic fertilizer, chemical pesticides, and mechanised irrigation systems contributed to yield increases. As activist Suman Sahai explains: "from that perspective it could not have been a better success; from a food importing, politically ambivalent India that was in an extremely difficult situation internationally [...]. This India was able to become independent of imports" (interview Sahai 2012). To the proponents, transgenic seed technology therefore are merely a continuation of this historical development-agricultural technology acquires the meaning of modernising agriculture, thus making India agriculturally self-sufficient and politically sovereign. As C. K. Rao from the pro biotechnology advocacy organisation Foundation for Biotechnology Awareness and Education (FBAE) contends: "India urgently needs improved technologies to develop quality seeds [...]. The Green Revolution demonstrated that various barriers to food production can be overcome through technological intervention" (Rao 2013, p. 171). To Rao, the whole history of domesticating plants is itself a history of "genetic modification only" (interview Rao 2013). This perspective regards GM crops not as different, but complementary to the agricultural technologies of the Green Revolution.

The sociotechnical imaginaries of productivity and environmentalism therefore suggest a linear continuation of the technological culture of the Green Revolution, based on the assumption of a simplistic causality that suggests technological progress equals progress of the human condition, from hunger to economic growth and environmental protection. A representative of Maharashtra Hybrid Seeds Company (Mahyco)[6] invoked the strong symbolism of the Green Revolution explicitly at a conference organised by the Indian state and industry organisations: "GM crops are the answer to the need for a second Green Revolution to increase food production and to cope with demands of population

---

[5]The term Green Revolution refers to the period after World War II in which improved agricultural technologies, such as high yielding crop varieties, inorganic fertilizers, chemical pesticides, and mechanized irrigation systems were applied to the agricultural systems of developing countries. These innovations led to productivity increases and made some countries from net importers to net exporters of food.

[6]Mahyco is India's largest seed company, having an approximated market share of 10% in crop seeds. Monsanto India Ltd, a subsidiary of Monsanto Company, USA owns 26% of Mahyco. Mahyco and Monsanto India together form the 50/50 joint venture Mahyco Monsanto Biotech. Mahyco distributes several Bt-cotton varieties and was centrally involved in the development of Bt-brinjal.

growth". Activist Suman Sahai explains the long-lasting impact of the Green Revolution on Indian culture: "Industry is not stupid, it does not say 'biotechnology', it says 'Second Green Revolution'; it does not say 'genetic engineering', but 'evergreen revolution'. It uses the same words that evoke the same images and reinforces the same positive believes" (interview Sahai 2012).

As Brooks (2005) explains, the narrative around a 'Second Green Revolution' enables the combination of contradictory discursive elements: It presents transgenic crops as natural, unavoidable, logical, and more sustainable while at the same time heralding them as the pinnacle of technological progress for agriculture. The imaginary of environmentalism links to the history of the Green Revolution differently. As the ABLE representative suggests, when it comes to risks to the environment, such as negative impact on biodiversity, "the Green Revolution is the culprit, not GM technology" (interview Seetharama 2013). Studies of Monsanto Company's business strategy suggest that representations of GM crops as solving environmental problems and addressing food security issues are the outcome of the conscious sales narrative that managers used to justify investments into biotechnology when the industry transitioned from producing agricultural chemicals to developing life-science based technologies (Glover 2010; Schurman and Munro 2010). In some sense the imaginary of environmentalism is a reply to the growing public demands for industrial organisations to produce technologies that are more environmentally friendly. The opponents of transgenic crops however, share a different interpretation of the history of the Green Revolution in India, and they accordingly construct different futures, to which I turn next.

2.3   Complex Ecology

The opponents of transgenic crops equally construct imaginaries of a sociotechnical future, though to them, transgenic seeds have an entirely different meaning. Civil society activists and NGO representatives question the positive visions constructed in the imaginaries above. Instead, the opponents argue that GM crops are part of a horrific future. Internationally renowned Indian activist Vandana Shiva[7]

---

[7]Shiva has published on many issues of agriculture (Shiva 1991), livelihood and ecology (Shiva 1988), and on GM crops (Shiva et al. 2000). Some describe her as a "rock star of the global fight against biotechnology" (Specter 2014). While many Indian grassroots level activists question the legitimacy of her campaign as representative of India's civil society and farming community, internationally she is often represented as "the voice leading the crusade against GMOs" (Frankman and Weinberger 2014).

depicts the future with GMOs as one in which the ecosystem will be brought out of balance: "if that system was allowed to spread […]. A destroyed planet will give no food at all. Dead soils, disappearing waters, a totally chaotic climate, no seeds. It is a recipe for an absolute, not just disaster, but a recipe for human extinction" (Shiva 2014). Constructing such a future pertains to a worldview that regards the complex interactions of the ecosystem as insufficiently understood by modern agricultural science. In a report by a network of anti-GM organisations, Shiva contends: "the camouflaged description of transgenic crops hides many of the ecological impacts of genetically engineered crops. The illusion of sustainability is manufactured" (Shiva et al. 2011, p. 12).

Instead, activists repeatedly emphasised in interviews the manifold risks they see implicated with the use of transgenic crops. To them, the risks to the ecosystem are too uncertain to release the technology into open agricultural fields without knowing about its long-term ecological impacts. For instance, doctor Ravikanth from the Ashoka Trust for Research in Ecology and Environment (ATREE) explained to me (interview Ravikanth 2013) that there is too little scientific data on assessing the potential impact of GM crops on India's environment. The concern about environmental risks pertains for instance to insects developing resistance to the Bt-toxin, the impact of the toxin on non-target organisms, and questions of outcrossing, i.e. gene-flow from GM plants to its non-GM relatives (for scientific studies on these issues, see Dorhout and Rice 2010; Ranjith et al. 2010; Zhao et al. 2010).[8] Indeed, from within the STI of complex ecology, the agricultural field is seen as part of the larger ecosystem.

This view accordingly emphasises that agricultural practice needs to be seen as a holistic activity within its environment. Pest problems, such as addressed by Bt-crops cannot be seen in isolation. A grassroots activist working with farmers in the Eastern Indian state Odisha explains for instance that brinjal farming is concerned with many more factors that make cultivating this vegetable a complex activity within a highly interdependent ecosystem: "Bt-brinjal gets multiple pests. If you do proper management of your farm, then the pest attack will not be such a big problem as the companies are trying to tell us. You have to take care of many more things than the pests-soil, seeds, and the whole circle of growing crops."

---

[8]Non-target organisms are organisms in the field which do not reduce yield; they may include soil organisms, non-pest insects, birds, and other animals. Although GM crops seem to be more target-specific than conventional pesticide applications, non-target effects with ecological consequences cannot be principally ruled out.

To this activist, insect-resistant crops are a solution to an imagined problem: "It is like you create a medicine first and then you create a disease" (interview Sarangi 2012). From the view of complex ecology, it is specifically environmental risk assessment that should receive more scrutiny in regulatory risk assessment of GM crops. However, as anti-GM crop campaigner Suman Sahai explains for the case of India: "in this mad race for GM crops, these sectors of agricultural science [plant breeding, etymology, agronomy, etc. …] get neglected in favour of this chimera called GM technology" (interview Sahai 2012). A holistic perspective backed up by science is therefore far from being part of mainstream agricultural policy in India.[9]

The critique of GM crops as an unpredictable, large-scale intervention into the complex ecosystem that is based on simplistic and linear thinking about the technological mastery of nature gets opposed by acro-ecological approaches that emphasise the complex interactions and interdependencies of the ecosystem. This, the opponents of transgenic crops argue, provides also a different perspective on natural and social orders linked to the way agriculture is done. Farmer activists Sarangi asks for instance: "To take the future of organic agriculture, which empowerment does it contribute to famers and what does it take to mother earth?" (interview Sarangi 2012). This emphasises that the opponents' imaginaries not only construct futures related to nature and its complexity, but such futuristic visions also draw an imaginary of the political ordering of society. Since both, scientific-technical and social orders get co-produced in the debate about GM crops, let me describe next how the sociotechnical imaginary of seed sovereignty gets constructed by the opponents of GM crops. I will then explain how both imaginaries are rooted in an interpretation of India's agricultural history distinctly different from the one offered by the proponents of transgenic seeds.

2.4   Seed Sovereignty

Indian activists criticise the idea that problems of malnutrition and hunger are caused by a lack of productivity. Rather, they point to the socio-economic consequences of industrial farming, i.e. questions of access to and ownership of the

---

[9]A notable exception here is the state of Andhra Pradesh, where non-pesticidal management has been taken up by the state government to upscale this less chemical-intensive practice of dealing with insect pests (for an analysis of how NPM addresses issues of vulnerability in the farming community, see Quartz 2011).

means of agricultural production such as seeds and land: "The crisis of agriculture is not a crisis of productivity [...]. It is a question of equitable access to productive resources" (interview Sarangi 2012). The STI of seed sovereignty criticises GM crops as part of the techno-scientific project of neo-liberal globalisation that will eventually lead to a monopolisation of the seed market and thus to a loss of farmers' control over seeds. Vandana Shiva argues that GM crops are "threatening the very basis of our freedom to know what we eat and to choose what we eat [...] our seed freedom is at peril" (Shiva et al. 2011, p. 7). In this imaginary, GM crops acquire the meaning of constituting risks to the socio-economic and political sovereignty of India's farming community. The future they envision with GM crops is one in which multinational corporations reduce the availability of seeds by centralising markets with the aim of transforming the agricultural system towards large monocultures and economies of scale. Such a transition to industrial farming, they argue, will lead to the transformation of the social relations of rural labour. As one farmer activist explained to me, GM crops is a technology full of politics, which prescribes to organise agriculture according to the principles of the market. This renders the farmer a consumer of seed technology, rather than a sovereign owner of productive resources who can make cultivation choices autonomously by reproducing non-proprietary seeds for the coming season. The issue of farmer suicides in India has been frequently invoked to be illustrative of the agrarian crisis modern agriculture has triggered. Various civil society actors have linked farmer suicides to the introduction of Bt-cotton, claiming that adoption of GM crops drives farmers into debt because of high input prices and agronomic failure of the technology (Kuruganti 2012; Sharma 2011, 2014; Shiva et al. 2011, 2000). A number of television documentaries follow a similar line of argument, bringing the imaginary of such a grim future into public circulation (Peled 2011; Sainath and Bhatia 2009). While some suggest the link between GM crops and suicides is made too easily (Gruère and Sengupta 2011), others regard suicides as a sign of wider agrarian distress, due to the transformation to industrial style farming which produces social injustice and leads to a loss of alternative political and cultural imaginations in the farming community (Shah 2012).

Disregarding the exact cause of India's farmer suicides, the issue of agrarian distress points to the critique of GM crops as representing a model of globalisation that benefits the profit-interests of multinational corporations instead of the livelihoods of developing countries' farming communities. From this perspective, transgenic crops themselves constitute a risk, since they are just one symptom of the privatisation of agricultural resources which has a profound impact on the practice of agriculture, and farmers' lives. The opponents argue that once in the market, GM crops leave farmers no room for sovereignty over cultivation decisions because of intellectual property rights that keep them from developing and exchanging their own bio-resources outside the capitalist economy. Activist Sahai explains: "Now there is complete control. No small seed store left in the US or in Europe. Now, they imposed this model on the rest of the world [...] that is why we have seed patents through TRIPS and the WTO" (interview Sahai 2012).[10] This perspective assigns GM crops the meaning of a technology of domination and control. Yet, similar to the STI of complex ecology, the STI of seed sovereignty does not only articulate a negative future. Rather, the imaginary of an alternative agrarian economy, one that mobilises its resources and claims its sovereignty is thriving in small pockets, particularly amongst members of farmer unions who oppose getting victimised by the neoliberal reshaping of India's agricultural policies which support the idea of a corporate-driven Second Green Revolution.

In February 2013, at the inauguration of the farmers' school Amritha Bhoomi in the South Indian state of Karnataka, I encountered the imaginary of seed sovereignty as a vision of the future that foresees farmers to reaffirm sovereignty. It was the NGO representative Babloo Ganguly from Timbaktu Collective, who urged farmers to claim their autonomy not only over the seeds they use, but also over the terms on which they participate in the market. In the future Ganguly imagined in his speech to a group of about eighty farmer leaders, transnational corporations would not dictate the conditions of participating in the market, but farmers themselves would influence these terms by forming cooperatives. This would allow them to include considerations for "water, the soil, the seed, the forest, our lifestyles" into the pricing and marketing policies. This, Ganguly urged, would allow farmers to regain sovereignty over their productive resources and decisions, as a move to more autonomy, freedom, and control over their own future. In an interview, Ganguly insisted that if the farming community started to reorganise and redefine its relation to the market for agricultural produce, it would gain true sovereignty over seeds and their lives. From this perspective, GM crops in contrast are a symbol of a future in which the farming community will lose their freedom and autonomy (interview Ganguly 2013).

---

[10]TRIPS refers to the Agreement on Trade-Related Aspects of Intellectual Property Rights (TRIPS) as agreed between the member states of the World Trade Organisation (WTO) at the end of the Uruguay Round of the General Agreement on Tariffs and Trade (GATT) in 1994.

How are the sociotechnical imaginaries of complex ecology and seed sovereignty linked to the actor's interpretation of past agricultural development? The opponents put forward an entirely different interpretation of the Green Revolution which revolves around its ecological, social, and political consequences. Kavitha Kuruganti, convenor of the Indian-wide network of farming activists Alliance for Sustainable and Holistic Agriculture (ASHA) explains that the success story of the Green Revolution is not universally accepted: "Now it is crossroads time, we are at the point to make decisions based on the last 50 years of experience in agriculture" (interview Kuruganti 2012).[11] This perspective interprets the Green Revolution to have established a causalistic, linear thinking about agricultural technologies' potential to increase output, which disregards the many ecological and social consequences of such a technological transitions. This interpretation of the past regards GM crop technology to perpetuate the same logic with potentially more severe consequences for the environment and farmers' livelihoods: "GM is not entirely new, it is part of the agricultural system that got established with the Green Revolution", yet "ecosystem-based conceptions of agriculture have a much better understanding of this than the modern reductionist science of GM" (interview Ramanjaneyulu 2012). Accordingly, this interpretation of the past emphasises how the Green Revolution imposed an epistemic frame that excludes other ways of knowing and doing agriculture. Several other activists agree: the Green Revolution destroyed indigenous knowledge and enslaved farmers to the capitalist mode of production, and GM crops are a continuation of this process: "It is certainly an extension of the past. It is the same set of players who want to make money with the old and new technology" (interview Kuruganti 2013).

The imaginary of seed sovereignty therefore interprets GM crops as a technology of domination that deprives farmers of their choice over seeds, once multinationals have established their dominant position in the seed market. In this sense, the STI of seed sovereignty puts forth a distinct interpretation of the socio-economic risks of transgenic crops. Leo Saldanha from the Bangalore-based NGO Environment Support Group (ESG) explains that GM crops are part of those structures of economic globalisation and technological modernisation that reproduce the power relations inherited from colonialism: "The structures are the same. Capitalism has adopted colonial structures; it is imperialism"

---

[11]Activists like Suman Sahai are however well aware of the role of the Green Revolution narrative in contributing to increased food security and India's independence from food imports in the post-independence period.

(interview Saldanha 2012).[12] Yet, the future implied by the STI of seed sovereignty is not completely grim. It entails a positive vision, one in which farmers are not victimised by technologies of domination and in which they affirm agency to engage in agricultural practice and economic relations differently. The universalising reductionism that the Green Revolution and GM crops symbolise therefore is opposed by a vision of ecological diversity and the struggle for political and economic autonomy of the farming community. Anthropologist Shiv Visvanathan offers an explanation on the symbolic meaning of diversity in the Indian context: it does not only refer to genetic and ecological diversity, but to the multidimensional ways of understanding life and livelihood, and thus to our understanding of the lived experience of those who are supposed to employ modern agricultural technologies (Visvanathan 1997).

3   Temporality, Modernity, and STS

So far, I have described four different sociotechnical imaginaries at play in the Indian debate about transgenic crops. These imaginaries of the future conflict and compete over which one constitutes the dominant interpretative frame to understand the meaning of transgenic crop technology. I have also shown where these imaginaries come from, i.e. how they root in diverging interpretations of past agricultural development. The Green Revolution narrative is central in these temporal constructions. Despite its hegemonic meaning as a technological leap that elevated India towards economic prosperity and political sovereignty, it remains contested in particular for its effects on the ecosystem and the farming community's sovereignty. Employing sociotechnical imaginaries as an analytical concept has helped to shown how the actors' constructions of sociotechnical futures relate to their interpretations of the past in the Indian GM crops debate. Sociotechnical imaginaries and the futures they construct need to be understood in the temporality between past, present and future.

This temporal dimension reveals how competing, conflicting, and contested sociotechnical futures and their connection to interpretations of the past reflect the relevant social group's experience of present social reality. A temporal perspective, in turn, allows for different historical, political, and normative

---

[12]From this perspective, the Green Revolution acquires the meaning of having established a system of dominating land and agricultural resources to control developing countries; and GM crops are mere continuation thereof.

perspectives (Adam 2006; Sharma 2013). I argue that it is important to understand GM crops technology within its wider temporal context of agrarian change and its implications for the agricultural relations of production in India: GM crops might lead to a future in which the technology solves problems of environmental degradation and productivity; but the technology can equally be understood as an instrument of domination that prescribes reorganising agricultural practices and rural labour in a certain way. The futures that break with the past of the Green Revolution instead imply respecting ecological complexity and asserting sovereignty over the means of agricultural production.

Understanding such temporal relations in technological controversies alerts us to ideas about nature, society, and time ascribed to and reflected in visions of technological change. For instance, the STIs of productivity and environmentalism rest on a mechanistic conception of time and change. Their temporality naturalises hegemonic relations of industrial organisation and production. These have shaped agricultural technologies and practices with certain conceptions about the technological mastery of nature and society. However, the STIs of complex ecology and seed sovereignty point us to another perspective of time, i.e. cultural and ecological time frames, which span over longer periods than the temporality of industrial capitalism (Adam 2006). For instance, the STI of seed sovereignty emphasise the struggle of some actors in the Indian context to conceive of alternatives to the structures imposed upon them by the institutions of the market and capitalist techno-science.

Understanding temporality as constitutive of the various meaning constructions of technology that actors imagine to profoundly change our sociotechnical future widens our understanding of the dimensions of technological controversy, as I have done here in the case of the Indian debate about GM crops. A temporal perspective can open up the study of such debates to the longer time-spans of socio-cultural and ecological time that technological change is implicated with. The historian Fernand Braudel has shown how human society evolves over the *longue durée* in which not only material, but also mental, conceptual, and environmental structures shape the course of social development beyond the immediate awareness of the actors involved (Hall 1980). I suggest that sociotechnical imaginaries is not only a useful lens through which to analyse the temporal dimension of debates about techno-scientific development, but I argue that sociotechnical imaginaries are by definition temporal. Thus, for understanding how sociotechnical futures reflect and shape the present, we have to consider the entire temporal scale spanning from past, to present, and the future. This can be a helpful insight when techno-scientific futures play a role in technology assessment and policy-making.

Although many of the issues addressed within the respective STIs get framed in the language of technological risks, particularly the last section on seed sovereignty shows the boundaries of such a discourse to be limited. STS scholar Brain Wynne has previously argued that the concept of risk is insufficient to understand technological controversies (Wynne 2005); and studies of the GM crops debates in other parts of the world have shown wider issues of agrarian change, agricultural systems, and globalisation to be implicated with technological risks (Brooks 2005; Heller and Escobar 2003; Jackson 2010). Although techno-science and its risks have fundamentally shaped modern society and culture (Beck 1992; Giddens 1990), I will not repeat the extensively debated limitations of the risk frame here. Rather, I want to point to the importance of studying non-Western empirical contexts, where modernity is experienced differently from the cultural, historical, and political contingencies of advanced capitalist economies (Appadurai 1996; Kaviraj 2011). For instance, Indian debates about social change are marked by a high degree of heterodoxy and a lively democratic discourse, from which Western democracies might learn (Sen 2005). In the case of the conflicting, competing, and contested futures in India's GM crops debate, I have shown temporal dimensions to be at work. Scrutinising debates about technological development with such a perspective that accounts for various time scales which operate at different levels, durations, and speeds can help to better understand the cause of controversy about new and emerging technologies in other places, too.

Since the construction of sociotechnical futures is as much about technical as about social change, studies of technological controversies can benefit from engaging more closely with temporality, and with the wider issues such temporal relations imply. Philosopher Charles Taylor describes the social imaginary as "ways people imagine their social existence, how they fit together with others, how things go on between them and their fellows, the expectations that are normally met, and the deeper normative notions and images that underlie these expectations" (Taylor 2004, p. 23). For Taylor, the social imaginary is part of economic and political modernisation: How the economy impacts people's life, how they constitute themselves as members of not only a social, but also a political sphere. In short, how they understand the technological projects around them as constituting normative frames, as enabling a sense of community, and as embodying conceptions of modernity. In that sense, understanding the temporality through the lens of sociotechnical imaginaries in debates about technoscientific change can enrich scholarly analysis and political practice alike. For instance, anticipatory technology assessment could accommodate temporal dimensions when drafting future scenarios for participatory technology assessment exercises; and policy-makers could accordingly get a better grasp of what technological change implies for the publics in whose interest they make decisions.

References

Adam, B. (2006). Time. *Theory, culture, society, 23*(2–3), 119–138.

Appadurai, A. (1996). *Modernity at large. Cultural dimensions of globalisation*. Minneapolis: University of Minnesota Press.

Beck, U. (1992). *Risk society: Towards a new modernity* (Trans. M. Ritter). London: Sage.

Bijker, W. E. (1995). *Of bicycles, bakelites, and bulbs: Toward a theory of sociotechnical change*. Cambridge: MIT Press.

Brooks, S. (2005). Biotechnology and the politics of truth: From the green revolution to an evergreen revolution. *Sociologia Ruralis, 45*(4), 360–379.

Carpenter, J. E. (2011). Impact of GM crops on biodiversity. *GM Crops, 2*(1), 7–23. https://doi.org/10.4161/gmcr.2.1.15086.

Choudhary, B., & Gaur, K. (2012). Socio-economic and farm level impacts of Bt cotton in India 2002–2010. http://www.isaaa.org/india/media/Socio-economic%20and%20farm%20level%20impact%20of%20Bt%20cotton%20in%20India,%202002%20to%202010-11%20aug%20final.pdf.

Collins, H. M. (1983). The sociology of scientific knowledge: Studies of contemporary science. *Annual Review of Sociology, 9*, 265–285.

CropLifeInternational. (2014). CropLife international-About. http://croplifefoundation.org/about/funders/.

Dorhout, D. L., & Rice, M. E. (2010). Intraguild competition and enhanced survival of Western Beat Cutworm (Lepidoptera Noctuidae) on Transgenic Cry1Ab (MON 810) Bacillus Thuringiensis Corn. *Journal of Economic Entomology, 103*, 54–62.

Express, T. I. (29 Aug. 2013). Seeds of change. *The Indian Express*. http://archive.indianexpress.com/news/seeds-of-change/1161406/.

Frankman, E., & Weinberger, J. (2014). Vandana Shiva, voice of the Anti-GMO debate. *The Take Away*. http://www.thetakeaway.org/story/vandana-shiva-voice-anti-gmo-debate/.

Giddens, A. (1990). *The consequences of modernity*. London: Polity Press.

Glover, D. (2010). Is Bt cotton a pro-poor technology? A review and critique of the empirical record. *Journal of Agrarian Change, 10*(4), 482–509. https://doi.org/10.1111/j.1471-0366.2010.00283.x.

Grant, H. (Ed.). (2008). Monsanto CEO grant discusses impact of genetically engineered crops on human health. http://www.eenews.net/tv/videos/874. Accessed 28 Aug 2008.

Gruère, G., & Sengupta, D. (2011). Bt cotton and farmer suicides in India: An evidence-based assessment. *Journal of Development Studies, 47*(2), 316–337.

Hajer, M. (1995). *The politics of environmental discourse: Ecological modernization and the policy process*. Oxford: Clarendon Press.

Hall, J. R. (1980). The time of history and the history of time. *History and Theory, 19*(2), 113–131.

Hammond, E. (2010). Counting the costs of genetic engineering. http://www.greenpeace. org/international/Global/international/planet-2/report/2010/1/counting-the-costs-of-genetic.pdf.

Heller, C., & Escobar, A. (2003). From pure genes to GMOs. Transnational gene landscapes in the biodioversity and transgenic food networks. In A. Goodman, D. Heath, & M. S. Linde (Eds.), *Genetic nature/culture. Anthropology and science beyond the two culture divide*. Berkley: University of California Press.

Jackson, M. (2010). Biotechnology and the critique of globalisation. *Ethnos: Journal of Anthropology, 67*(2), 141–154.

Jasanoff, S. (Ed.). (2004). *States of knowledge: The co-production of science and social order*. New York: Routledge.

Jasanoff, S., & Kim, S. H. (2009). Containing the atom: Sociotechnical imaginaries and nuclear power in the United States and South Korea. *Minerva, 47*, 119–146.

Jasanoff, S., & Kim, S. H. (2013). Sociotechnical imaginaries and national energy policies. *Science as Culture, 22*(2), 189–196.

Jasanoff, S., & Kim, S. H. (2015). *Dreamscapes of modernity: Sociotechnical imagianries and the fabrication of power*. Chicago: The University of Chicago Press.

Kaviraj, S. (2011). *Introduction to the enchantment of democracy and India. The enchantment of democracy and India* (pp. 1–24). Bangalore: Permanent Black.

Kuruganti, K. (2012). Bt cotton, a bitter havest for farmers: Suicide and despair in India. http://climate-connections.org/2012/06/02/bt-cotton-a-bitter-harvest-for/.

Marcus, G. E. (Ed.). (1995). *Technoscientific imaginaries: Conversations, profiles, and memories*. Chicago: Chicago University Press.

Monsanto. (2012). Sustainability report. http://www.monsanto.com/sitecollectiondocuments/csr_reports/2012-csr.pdf.

Padmanaban, G. (2013, September 2). Sow the wind, reap a storm, opinion. *The Hindu*. http://www.thehindu.com/opinion/op-ed/sow-the-wind-reap-a-storm/article5082915. ece.

Peled, M. X. (2011). *Bitter seeds. An examination of the debate surrounding biotechnology and the future of farming*. San Francisco: ITVS. https://itvs.org/films/bitter-seeds.

Pinch, T. J., & Bijker, W. E. (1984). The social construction of facts and artefacts: Or how the sociology of science and the sociology of technology might benefit each other. *Social Studies of Science, 14*, 399–441.

Pinch, T. J., & Bijker, W. E. (1987). The social construction of facts and artifacts: Or how the sociology of science and the sociology of technology might benefit each other. In W. E. Bijker, T. P. Hughes, & T. Pinch (Eds.), *The social construction of technological systems. New direction in the sociology of technology* (pp. 17–50). Cambridge: MIT Press.

Powerbase. (2014). Powerbase public interest investigations. Profile: CropLife international. http://powerbase.info/index.php/CropLife_International.

Pritchard, B., Rammohan, A., & Sekher, M. (2013). Food security as a lagging component of India's human development: A function of interacting entitlement failures. *South Asia-Journal of South Asian Studies, 36*(2), 213–228. https://doi.org/10.1080/0085640 1.2012.739256.

Quartz, J. (2011). *Constructing Agrarian alternatives. How a creative dissent project engages with the vulnerable livelihood conditions of marginal farmers in South India*. Maastricht: Universitaire Pers Maastricht.

Ranjith, M. T., Prabhuraj, A., & Srinivasa, Y. B. (2010). Survival and reproduction of natural populations of Helicoverpa Armigera on Bt-cotton hybrids in Raichur, India. *Current Science, 99*(11), 1602–1606.

Rao, C. K. (2013). Genetically engineered crops would ensure food security in India. In D. J. Bennett & R. C. Jennings (Eds.), *Successful agricultural innovation in emerging economies. New genetic technologies for global food production*. Cambridge: Cambridge University Press.

Roulston, K. (2010). Considering quality in qualitative interviewing. *Qualitative Research, 10*(2), 199–228.

Sainath, P., & Bhatia, D. (2009). Nero's guests. Amsterdam & New Delhi. https://www.idfa.nl/en/film/e9265992-0425-4592-acca-29d2ba61010b/neros-guests.

Schurman, R., & Munro, W. A. (2010). *Fighting for the future of food. Activists versus agribusiness in the struggle over biotechnology*. Minneapolis: University of Minnesota Press.

Sen, A. (2005). *The argumentative Indian. Writings on Indian culture, history and identity*. London: Penguin.

Shah, E. (2005). Local and global elites join hands: Development and diffusion of genetically modified Bt cotton technology in Gujarat. *Economic and Political Weekly, 40*(43), 4629+4631–4639.

Shah, E. (2012). A life wasted making dust: Affective histories of dearth, death, and debt and farmers' suicides in India. *Journal of Peasant Studies, 39*(1), 1–21.

Shapin, S., & Schaffer, S. (1985). *Leviathan and the air-pump: Hobbes, boyle and the experimental life*. Princeton: Princeton University Press.

Sharma, D. (2011, January 19). Ground reality: Killer technologies will not increase our food production. http://devinder-sharma.blogspot.nl/2011/01/killer-technologies-will-not-increase.html?q=killer+technologies.

Sharma, D. (2014, February 6). Ground reality: Prime Minister ignores the facts. Openly bats for dangerously risky GM crop technology. http://devinder-sharma.blogspot.nl/2014/02/prime-minister-ignores-facts-openly.html.

Sharma, S. (2013). Critical rime. *Communication and Critical/Cultural Studies, 10*(2–3), 312–318.

Shiva, V. (1988). *Staying alive: Women, ecology and survival in India*. New Delhi: Kali for Women.

Shiva, V. (1991). *The violence of the green revolution*. London: Zed Books.

Shiva, V. (Ed.). (2014, Febrary 21). Transcript speech at food otherwise conference. http://www.voedselanders.nl/voedselanders.nl/Start_files/FINAL%20transcript%20vandana%20shiva-%20plain%20text.pdf.

Shiva, V., Barker, D., & Lockhart, C. (2011). The GMO emperor has no clothes: A global citizen's report on the state of GMOs-False promises, failed technologies. https://www.nabu.de/imperia/md/content/nabude/gentechnik/studien/gmo_emperor_study_pdf.pdf.

Shiva, V., Jafri, A. H., Emani, A., & Pande, M. (2000). *Seeds of suicide: The ecological and human costs of globalisation of agriculture*. New Delhi: Research Foundation for Science, Technology and Ecology.

Smith, E. (2009). Imaginaries of development: The rockefeller foundation and rice research. *Science as Culture, 18*(4), 461–482.

Specter, M. (2014, August 25). Seeds of doubt. An activist's controversial crusade against genetically modified crops. *The New Yorker*. http://www.newyorker.com/magazine/2014/08/25/seeds-of-doubt.

Stone, G. D. (2012). Constructing facts. Bt cotton narratives in India. *Economic and Political Weekly, XLVII*(38), 62–70.

Taylor, C. (2004). *Modern social imaginaries*. Durham: Duke University Press.

Visvanathan, S. (1997). *Footnotes to Vavilov: An essay on gene diversity a carnival for science: Essays on science, technology, and development*. Delhi & New York: Oxford University Press.

Wynne, B. (2005). Risk as globalising democratic discourse? Framing subjects and citizens. In M. Leach, I. Scoones, & B. Wynne (Eds.), *Science and citizens: Globalisation and the challenge of engagement* (pp. 66–82). London: Zed Books.

Zhao, J. H., Ho, P., & Azadi, H. (2010). Benefits of Bt cotton counterbalanced by secondary pests? Perceptions of ecological change in China. *Environmental Monitoring and Assessment, 173*(1–4), 985–994.

List of Semi-Structured Interviews

Ganguly, B., activist Timbaktu Collective, conducted at Amritha Bhoomi, Chamaraja Nagar District, Karnataka, 14.02.2013.

Kuruganti, K., Convenor Alliance for Sustainable and Holistic Agriculture (ASHA), conducted in Bangalore, 16.02.2012.

Kuruganti, K., Convenor Alliance for Sustainable and Holistic Agriculture (ASHA), conducted in Bangalore, 08.01.2013.

Padmanaban, G., molecular scientists Indian Institute of Science, conducted in Bangalore, 19.02.2012.

Ramanjaneyulu, G. V., Executive Director Centre for Sustainable Agriculture (CSA), conducted in Hyderabad, 27.03.2012.

Rao, C. K. General Secretary Foundation for Agricultural Biotechnology and Awareness (FBAE) conducted in Bangalore, 10.01.2013.

Ravikanth, G. scientist Ashoka Trust for Research in Ecology and Environment, conducted in Bangalore, 07.02.2013.

Sahai, S., Director Gene Campaign, conducted in Delhi, 28.02.2012.

Saldanha, L., Coordinator Environment Support Group (ESG), conducted in Bangalore, 20.02.2012.

Sarangi, activist Living Farms, conducted in Bhubaneswar, 17.03.2012.

Seetharama, N., Executive Director Association of Biotech-Led Enterprises (ABLE), conducted in Delhi, 28.01.2013.

Andreas Mitzschke (Ph.D.) is post-doctoral researcher and teacher in Science, Technology & Society Studies at the Faculty of Arts and Social Sciences at the University of Maastricht.

Smart City Experimentation in Urban Mobility—Exploring the Politics of Futuring in Hamburg

Philipp Späth and Jörg Knieling

Abstract

This chapter explores how a Smart City agenda has influenced attitudes towards the future in contemporary mobility planning in Hamburg. By comparing three recent frameworks of transportation planning, we detect an interesting shift that occurred when Hamburg's administration embarked on the project of becoming a leading Smart City. At that point in time, an attitude of planning, characterized by the styles of foresight and prediction, by practices of calculating and by the logic of precaution was replaced, or at least complemented and challenged, by an attitude of experimenting towards real-time management, which is characterized by a style of premediation, practices of performing and a logic of preparedness (in terms of Anderson 2010). We discuss multiple implications of such a shift on governance arrangements and prospects for citizen participation and decision making.

P. Späth (✉)
Institute of Environmental Social Sciences and Geography,
University of Freiburg, Freiburg, Germany
e-mail: spaeth@envgov.uni-freiburg.de

J. Knieling
Institute of Urban Planning and Regional Development,
HafenCity Universität Hamburg, Hamburg, Germany
e-mail: joerg.knieling@hcu-hamburg.de

© Springer Fachmedien Wiesbaden GmbH, part of Springer Nature 2019   161
A. Lösch et al. (eds.), *Socio-Technical Futures Shaping the Present*,
Technikzukünfte, Wissenschaft und Gesellschaft / Futures of Technology,
Science and Society, https://doi.org/10.1007/978-3-658-27155-8_8

1 Introduction

The term 'Smart City' has gained increasing prominence in debates about visions and guidelines for future urban development (de Jong et al. 2015; Mora et al. 2017). Under this notion, at least in larger cities in Europe, Asia and North America, nearly all stakeholders engaged in urban governance occasionally, and in many cases very frequently, refer to the benefits or risks of fundamental, ongoing or potential transformations of urban infrastructures and urban life related to the implementation of innovative information and communication technologies in various fields of urban development.

In the early days of this development, around 2011 to 2014, many scholars and organizations tried to establish tangible definitions of what makes a city smart (Caragliu et al. 2011; Albino et al. 2015). Lately, more and more scholars and practitioners of urban governance view such attempts as futile or even undesirable. This holds particularly for countries like Germany, where municipalities and regions enjoy a relatively high degree of political importance and autonomy, and hence where the definition of what would make a particular city or region "smart" is commonly left to 'local' actors in the respective places (Spaeth et al. 2017). At the same time, the term 'Smart City' does convey certain meanings, even if these are vague. It regularly triggers associations of an increasing interconnectedness between different sectors and urban infrastructures (Song et al. 2017; Kourtit et al. 2012), and it is commonly associated with the use of information and communication technologies, particularly newly available data from sensors, mobile devices and social media for the management and planning of urban infrastructures. In this sense, the notion of 'Smart City' signifies a particular vision of urban (infrastructure) development: interconnected, based on (big) data, and leading to improved services and liveability (Goldsmith and Crawford 2014; Townsend 2013). A key storyline that is centrally positioned in all positive expectations regarding the smartification of cities and regions is that the use of information technology and data would enable an increase in the effectivity and efficiency with which energy, water, mobility and other services are provided. At the same time, many scholars and practitioners are concerned about possible side effects of the emerging technologies and socio-technical configurations: the vulnerability of crucial, but increasingly interconnected infrastructures (Vanolo 2015); threats to privacy; and the potential for a so far unimagined surveillance of people (Sennett 2012; Kitchin 2014; Vanolo 2014). Being a shortcut to utopian, dystopian or ambivalent visions of usually vaguely sketched socio-technical futures in urban environments, the 'Smart City' notion increasingly transcends the sphere of urban infrastructure development. Today, it is increasingly recognized that "Smart City" initiatives are inherently multi-dimensional, and do not only have an impact on technology and material infrastructures, but also a crucial influence on the development of urban governance arrangements, the relationship between public and private actors, between citizens and state authorities and amongst citizens.

Whether they are called 'Smart City initiatives' explicitly, 'digitization strategies' (as many in Germany), or something else, initiatives that relate to the Smart City vision cannot be studied from a discursive perspective only. Many attempts to make cities smart will, at least in the long run, also involve quite tangible, material changes to infrastructures. Furthermore, a broad range of institutions—i.e. rules of different kinds—are often created or modified where the Smart City ideal is taken up.

In this chapter, we aim to understand how the Smart City ideal, as a sociotechnical future, that floats around in many different but globally connected forms, can influence contemporary debates on the future in a particular place. Using the German city-state of Hamburg—and the way in which alternative futures of its transportation system have been negotiated in recent years—as an illustrative example, we seek to reconstruct how imagined future socio-technical developments resulted in a discursive shift regarding an appropriate attitude towards the future: How future challenges and opportunities should be anticipated, by whom, and what kind of preparations for the future seem possible and necessary.

Considering the analytical framework that was developed for this book, we ask:

1. What approach to anticipating the future is proposed in three recent policy frameworks that seek to prepare for contemporary and future challenges in Hamburg's transportation infrastructure?
2. Who is promoted in these three contexts as key actors in preparing for the future?
3. How do these varying attitudes to futuring affect the present and, in particular, contemporary transportation planning practices?

In the following, we first introduce a conceptual approach (following Andersen 2010) for unpacking "styles, practices und logics" that reveal the multiform presence of the future in contemporary attitudes, practices and geographies (Sect. 2). This is followed by a brief introduction of our methods (Sect. 3). We then present the findings related to three recent policy frameworks that seek to prepare for contemporary and future challenges in Hamburg's transportation infrastructure: the Mobility Program of 2013, the ITS-Hamburg initiative from 2016 and the mySMARTlife project that was also started in 2016 (Sect. 4). The latter two frameworks were both developed explicitly to position Hamburg as an international front-runner in the smartification of urban (transportation) infrastructures. Taken together, we argue that they illustrate how smart city initiatives can, in the space of very few years, comprehensively change attitudes with regard to anticipating future challenges and opportunities for transport infrastructures (Sect. 5). Some conclusions in Sect. 6 finally lead us back to the initial research questions and reflect on an emerging research agenda.

2 Conceptual Approach

We follow the call to study an "actually existing" smart city and

> "how relationships within the city are changing, especially with respect to ways of imagining the different spaces of the city and the 'urban problems' posed by and within such spaces, and what kinds of interventions might be designed to ameliorate these problems. In order to better understand the geographically-differentiated spaces of the smart city, it is important that we ask how visions of the data-driven, smart city are actually playing out in specific cities and neighbourhoods." (Shelton et al. 2015, p. 17)

In our attempts to answer the research questions above, we can build on some long-lasting traditions within science and technology studies to explore how socio-technological developments are influenced by certain images of the future (van Lente 1993; Brown et al. 2000). This influence has been found to be constrained by many factors. Technological development is usually a highly distributed and complex process shaped by largely unpredictable influences including relatively free decisions of many human agents involved. Nevertheless, like many other authors (Grunwald 2011, 2012; Jasanoff 2015; Raven 2017), we see a great potential in studying those images of the future that find strong resonance within society or at least among those groups of professionals whose decisions and actions can influence how socio-technological arrangements are shaped in the present. In our case, we study how ideas of a smart or digitized City of Hamburg shape the ways in which the city and its infrastructures are transformed in the present.

With regard to a "global imaginary" on one hand as well as "situated instances" of the smart city on the other hand, White (2016, p. 585) observed a very particular approach to the future: "Rather than rely on affirmative action toward some normative goal, the smart city's global imaginary resorts to the threat of inaction spiralling into panic and insecurity. This post-modernist turn in urban planning and policy relies not so much on a politics of optimism, as on one of realistic-seeming fear, tempered by the possibility of hope—but only if action is taken now." (White 2016, p. 584)

With our contextualized research of instances of a smart city debate in Hamburg, we would like to explore the extent to which local experiments in transportation reflect this assumed shift.

In order to structure our analysis—particularly with regard to our first research question on how the approach to anticipating the future has been changing in line with the increasing prominence of a smart city discourse amongst many actors in Hamburg—we will follow Anderson (2010) and examine three elements of an attitude towards the future: "*styles* through which the form of the future is disclosed and related to; *practices* that render specific futures present; and *logics* through which anticipatory action is legitimized, guided and enacted" (Anderson 2010, p. 777, emphasis using italics added).

Regarding *styles* of disclosing futures, Anderson distinguishes (a) "foresight based on good judgement" from (b) "probabilistic prediction based on induction from the past distribution of events". Moreover, he considers that a third style (c) which he calls "premediation" is increasingly gaining importance: This style is based in "possibilistic thinking" and a "disclosing of the future as a surprise" (Anderson 2010, p. 782). Following Anderson, we can also distinguish three main *practices* of rendering futures (that are increasingly anticipated to come as a surprise) actionable. These are practices of (a) calculating and modelling, (b) imagining and visualizing, and (c) performing, experiencing and thus embodying possible futures (Anderson 2010, pp. 783–787). Finally, *logics* of precaution, pre-emption and preparedness can be distinguished based on the different extents to which action is legitimized in the absence of clues about which futures are actually possible or impossible and what the implications of particular undesired events might be (Anderson 2010, pp. 787–792).

All this is meant to acknowledge "the presence of the future" (Anderson 2010, p. 783); meaning that socially dominant practices of anticipating the future shape socio-spatial relationships (alias geographies) in the present.

Many routines of urban governance are currently being reconsidered due to the fact that sensors and actuators are nowadays cheaply available which can communicate in an 'internet of things' (IoT) and together build up a comprehensive database for the shaping and management of infrastructures that was previously unimaginable (Zook 2017). We will therefore examine how the currently debated (future) smartification of urban transportation[1] infrastructures may transform, for example, the ways in which governments and citizens interact with each other (Vanolo 2016) or what (new) ways of governing and managing urban infrastructures are considered appropriate (Meijer 2017).

At the same time that urban governance was increasingly focusing on the promises and risks of digitization, a paradigmatic shift occurred in terms of what were considered timely and appropriate forms of collective action regarding the development of urban infrastructures, at least in Europe. While urban governance in the past largely followed the ideal of planning—seen as a systematic process of deliberation and decision making based on the anticipation of future demands and challenges—we now see a near ubiquitous trend that decisions should or can be taken only after some "experimentation" (Evans et al. 2016). As the widely influential idea and practice of creating "urban living labs" shows, this is particularly prominent in relation to urban governance and in the context of the ambition to achieve urban sustainability (Bulkeley and Castán Broto 2013; Bulkeley et al. 2016; Sengers et al. 2016; Sengers et al. 2018).

Problems in urban mobility are commonly seen as a primary field of application for the newly available digital technologies and big data (Schwanen 2017). At the same time, the very fundamental shifts that are anticipated in this field (electrification, automated driving, marginalization of vehicle ownership etc.) and the openness of future developments are considered a strong reason for following an experimental approach which is often portrayed as being more cautionary compared to direct investments based on planning (Berliner Erklärung 2017). We therefore find it very promising to explore how these two developments—the focusing on digitalization and the turn to experimentation as a mode of governance—have influenced the governance of transportation infrastructure in Hamburg.

---

[1]The authors acknowledge that there is a crucial difference between a perspective on (means and flows of) transportation and on (needs and experiences of) mobility, especially from a planning and urban development perspective. The documents and actors that the contribution refers to usually speak of means/infrastructures/policies/planning of "transportation" (in German: "Verkehr") and that is reflected in this chapter. Nevertheless, the authors adopt a broader perspective on mobility (cf. UN-Habitat 2013) wherever possible.

3 Methods

There are no well-established conceptual and methodological approaches to studying the subject in which we are interested. We explore these developments in the form of three illustrative case studies of transportation planning and Smart City experiments in the field of mobility in Hamburg, Germany. To guide our analysis, we specify our guiding questions with a focus on different styles, practices and logics of anticipatory action as distinguished by Anderson (2010).

In this endeavour, we can generally build on work that has recently been conducted by Philipp Späth and colleagues in the DFG-funded research projects Smart-Eco-Cities.org and SmartKnowledgePolitics.com as well as by the EU project consortium for "mySMARTlife" that Jörg Knieling is part of. He can furthermore draw on observations made during many years of actively observing urban politics in Hamburg, e.g. in his role as a member of the "Dialogue Board" of the Ministry of Urban Development and as a researcher in a number of Hamburg related projects.

Empirically, the case studies below are based on an analysis of published material (policy documents, working papers, speech manuscripts, websites, brochures etc.) and grey literature as well as about a dozen interviews that were conducted in 2015–2017 with knowledgeable people of different backgrounds (various departments of the city administration, publically and privately owned companies, the port authority, civic organizations, academia and journalism). We coded key policy documents and the transcripts of the interviews with the help of qualitative data analysis software, which helped us to single out quotes representing attitudes towards the future and the differences and similarities between the three programs.

4 Three Frameworks for Transportation Planning in Hamburg as a Window to Understanding the Politics of Futuring

With its roughly 1.8 million inhabitants, Hamburg is the second biggest city in Germany. The city region's territory and administration also serves as one of the sixteen German federal states. About 2.8 million people live in the agglomeration of Hamburg, which in turn forms part of the larger "Hamburg Metropolitan Region", home to about 5.3 million people. The city encompasses the port of Hamburg in which up to 140 million tons of goods are processed annually, making it the biggest harbour in Germany and the third largest in Europe. The city has traditionally been characterized as a particularly open, internationally visible one, and the current administration sees it competing with Shanghai and Sydney to be one of the most attractive, innovative and liveable cities by the sea.

Over the last years, several Eco and Smart City projects have been planned or implemented throughout the city (Spaeth et al. 2017). When a new Government was formed in early 2015, "digitization" was a key theme of the coalition agreement between the social democrats and the green party.[2] Just prior to this, the senate had passed a "Digital City Strategy" to bundle, influence and coordinate processes of digitalization in many fields, among them education and health. Consequently, a coordination office ("Leitstelle Digitale Stadt") was established to ensure that the governmental and private actors cooperate and that the comprehensive perspective of the municipality is acknowledged in these processes.[3]

Hamburg's lord mayor, Olaf Scholz, has repeatedly argued for an engagement of the state and municipal government in Smart City experiments to ensure that the unavoidably digitalized future is co-shaped with the public good in mind (cf. public speeches on 30 April 2014 and 2 May 2016). In line with this ambition, the municipality in its role as state government first engaged in bilateral and multilateral negotiations with system providers (IBM, Microsoft, Cisco etc.) and network operators (Vattenfall, Eon-Hanse etc.) to establish visible Smart City experiments in Hamburg. Between 2011 and 2014, various contracts and memorandums of understanding between the city of Hamburg and large companies were publically announced. These agreements loudly signalled the will to cooperatively develop Smart City pilot projects and applications in public-private partnerships. Since then, however, little information has been made available on how these particular pilot projects have developed and whether any lessons have been learned.

After the lord mayor was re-elected in 2015, the new coalition government of social democrats and now also the green party agreed on a digitalization strategy and institutionalized an officer for digitalization within its administration. Already since 2014, a series of roughly quarterly meetings have allowed for delegates from state departments, semi-public enterprises and private partners to coordinate their Smart City experiments in the city. At the beginning of 2018, a new

---

[2]At this occasion, a previous attempt to position Hamburg as Germany's primary "Smart City" (cf. MoU CISCO—Hamburg 2014) was twisted towards a "proactively digitized" city, allegedly because "Smart City" was expected to provoke negative associations in a German population and because of scepticism among local government officials themselves.

[3]This and the following two paragraphs have partly been derived from some earlier, comparative work by P. Späth and colleagues (Raven et al. 2017).

administrative structure ("Koordinierungsstelle Urban Data Hub") was built for the coordination of the integrative data infrastructure "Urban Platform Hamburg". From January 2018 on, the digitization activities in Hamburg are furthermore coordinated by a newly established CDO. The creation of this new, high ranking position was welcomed by our interviewees as a sign of Hamburg's ambition to lead in Smart Urbanism. After Hamburg's Lord Mayor Scholz—who has driven the digitization agenda since its early days—has accepted a position as Finance Minister in the new federal government of Germany in March 2018, this re-staffing of the digitization unit may contribute to ensuring that Hamburg's ambitions to lead in digitization will be sustained.

4.1   Comparing Three Frameworks for Planning Transportation Infrastructure in Hamburg

The idea of Smart Cities has now been around for many years. In Hamburg, this was reflected in some conferences and negotiations between large IT infrastructure and service providers which took place from 2011. However, it was only in 2014 that a first significant initiative was launched by the Lord Mayor (Spaeth et al. 2017). Just before this, in 2013, an official framework for transportation planning was approved by the Senate that was meant to guide the transportation infrastructure development in Hamburg through to the year 2020. Just after, in 2016, two other initiatives were developed, which have since strongly influenced the development of transportation infrastructure. Both ongoing projects are clearly imbedded in (internationally shaped) Smart City agendas and regularly mentioned as prime examples of Hamburg's leading role in proactive digitization: The department for economic development's initiative to develop a model Intelligent Transportation System (ITS) in Hamburg as well as the European funded Smart City Lighthouse project mySMARTlife. This presents a fortunate opportunity to study the ways in which the contemporary smart city experiments deviate in their approach to futuring from the only marginally older, incumbent approach to transportation planning in Hamburg.

4.1.1 The Mobility Program for 2014–2020

In September 2013—i.e. just before Hamburg was first positioned as a pioneer Smart City—the state government of Hamburg[4] decided on a "Mobility Program"

---

[4]In the three German city-states, the cabinet is called "Senate", and we report here that the strategy was adopted by the Senate.

which aimed to lay out a trajectory for transportation planning during the period 2014–2020 (Hamburger Senat 2013). The mobility program starts with a review of the available forecasts regarding future needs of transportation of goods and passengers (pp. 11–17), covering the period to 2020, with some outlooks for 2050. It describes the administration's preferred scenarios regarding urban growth (pp. 18–24) and discusses the current and future availability of financial resources (pp. 27–29). In Anderson's terms, the *styles* of anticipating the future that are propagated here are "foresight based on good judgement" and "probabilistic prediction based on induction from the past distribution of events" (Anderson 2010, p. 782). Against the background of such expertise-based predictions, the program then lays out a number of activities that the administration intends to pursue or could consider pursuing, and defines the actors, procedural principles and sequential steps of transportation planning for Hamburg. Defining these procedures, at least in such a comprehensive form, is a non-obligatory activity of the administration, as is stated on (p. 61).

Concerning the procedural prescriptions, the program contains a sequential strategic planning process with periodical updating. Building on a guiding vision ("Leitbild") and a list of objectives, analyses are conducted, followed by pre-set methods, often including scenario building exercises, before, finally, fields of activity and measures are agreed upon (cf. the graph on p. 60). For coordination between the multitude of involved administrative departments and the coordination of these with state-owned utilities and private providers of transportation services, as well as with the administration of hinterland areas, a specific advisory committee ("Mobilitätsbeirat") was established. The document announces that a simulation model of transportation needs and flows will be developed for long-term use, and that this development will be commissioned on the basis of a competitive tendering among invited specialists (p. 62). Developing such a quantitative tool for extrapolation of trends can clearly be subsumed to the "calculating futures" category of *practices* by Anderson (2010, p. 784). For the identification and prioritization of improvements in transportation infrastructures and mobility management activities, it lays out a multi-year schedule, emphasizing a clear sequence of phases that are each concluded by reaching a milestone: The identification of objectives and scenarios, through 2014 and 2015, was to be completed with a binding decision on the "framework of objectives" ("Zielrahmen"). A following phase of clarifications and decision making, through 2016 and 2017, was set to conclude with the adoption of an official "transportation development plan". The final phase of implementation, through 2017 to 2020, was envisaged to terminate with a last "implementation report" in 2020 (p. 63). The whole sequenced program is clearly motivated and characterized by the attitude of

"Vorsorge", or taking care before acting, which Anderson typified as the *logic* of "precaution" (2010, p. 788)

The program also clearly reflects an ambition to make transportation planning a rational, continuous and transparent process, in which public authorities lead multi-stakeholder consultation processes in a predefined, systematic process. A section on "transportation planning with whom" (p. 61 f.) defines exactly which public authorities (across governance levels) and which external partners have to be involved in order to ensure acceptance of the final decisions. In each step, and particularly in the phase of articulating political objectives, this process is clearly open to political contestation. Debates are foreseen primarily within the mobility advisory board, in which "the relevant groups of society" are to be represented (p. 61). To a certain extent, however, the making transparent of "objectives, targets, scenarios and measures" would also enable civil society organizations, which are not part of the advisory board, and even individual citizens, to voice concerns and preferences.

The advisory board has met 11 times since its establishment in 2014.[5] A set of objectives and target values was published in January 2017, but a comprehensive transport development plan (VEP), which was originally planned to be released by 2017 has not yet been adopted by the Senate (as of March 2018).

4.1.2 Intelligent Transportation Strategy (ITS) for Hamburg

The Intelligent Transportation Strategy for Hamburg (ITS-HH) is a political initiative to strategize and coordinate the build-up of top-of-the-range infrastructure and standards for data management and exchange. The strategy passed the Senate in April 2016, starting a two-year period for experimentation with technical solutions. For this testing phase, the Senate decided to invest 1.85 mio. Euro (of which 600.000 Euro were allocated to preparing a bid for hosting the ITS-world congress in 2021, which was successful in late 2017) (BWVI Hamburg 2016). In this strategy, the future is described as being open to an unprecedented extent. Therefore, experiments shall be conducted to inform future decisions about which technology and standards will be rolled out more broadly.

The strategy relates to the internationally used term ITS and defines it as "transportation infrastructure, ICT and vehicle systems that—by interconnecting vehicles, infrastructure and people (e.g. via mobile phones) and by an exchange of real-time data—contribute to making mobility more reliable, safe, efficient and

---

[5]See http://www.hamburg.de/bwvi/mobilitaetsbeirat/.

environmentally sound" (BWVI Hamburg 2016, p. 5, translated by the authors). Accordingly, the objectives of the ITS-HH are to "improve transport safety; reduce environmental impacts of transportation; improve reliability and efficiency; support good and safe data collection and exchange of information, foster innovations" (BWVI Hamburg 2016, p. 15).

The strategy builds on the claim that recent technological developments induce fundamental changes in the transport sector: "Currently, a developmental leap is going on due to the megatrends of IT: internet-of-things, mobility, cloud, big data and social network" (BWVI Hamburg 2016, p. 5). The document reviews where elements of an ITS are already in place in Hamburg and identifies eight fields of action. "Data" ("quality/interoperability/safety, authorized management" BWVI Hamburg 2016, p. 18) and "innovation" (ITS projects will "create a climate that is encouraging experimentation and innovation", BWVI Hamburg 2016, p. 24) are defined and visualized as cross-cutting issues. "Information, intelligent traffic control, intelligent infrastructure, intelligent parking, mobility as a service and intelligent vehicles" (BWVI Hamburg 2016, p. 18) are depicted as six columns that will structure the work of the diverse actors to address the questions that are repeatedly arising in all fields: What kinds of data are available, which are needed? [...] What level of automation can be applied (e.g. in traffic control)? What strategies and defaults is this based on? [...] Which technology is appropriate and in compliance with data security/privacy regulation? [... and] How can the long-term utility of investments be ensured (e.g. via non-proprietary solutions)?" (BWVI Hamburg 2016, p. 24 f.). It is envisaged that addressing these questions (within two years from April 2016) will lay the groundwork for an optimally future-proof infrastructural basis for ITS in Hamburg (BWVI Hamburg 2016, p. 25).

The unprecedented uncertainty and ever-shorter product lifespans characterizing today's technological development are mobilized in support of an experimental and cooperative approach to the governance of infrastructural investments. A key element of this approach is that future technologies and forms of cooperation (based e.g. on an exchange of data) need to be tested in small 'pilots' with the involvement of all relevant stakeholders (i.e. business, public authorities and research institutions) who each bring in their diverse views and interests and increasingly learn how to cooperate in order to achieve the common goals (which are vaguely described in the ITS strategy).

The strategy consequently stresses an experimental approach of "piloting innovative developments in prototypes and small projects and, in case they stand the test, employing them more broadly in standard applications" (BWVI Hamburg 2016, p. 27). Following this quasi-evolutionary approach and increasingly cooperating with "innovative companies" is furthermore expected to "create a climate of joyful experimentation and innovation" and to "strengthen Hamburg's role as a focal point of development in the field of ITS" (BWVI Hamburg 2016, p. 24).

Yet, the experimentation is given a broadly defined timeframe ("five to ten years"), a systematic, sequential process structure and is oriented towards specific goals (that should stay valid "at least until 2022", BWVI Hamburg 2016, p. 28) by means of the strategy document that was developed under the leadership of the department for economic affairs and transportation (BWVI Hamburg 2016). A first report on a proposed ITS framework architecture has been promised for mid-2018, but this will be preliminary, requiring constant updating thereafter. In late 2017, key experts involved in the development of this report still considered major technological developments to be very open, unpredictable and too immature to allow for large investments. At the same time, a huge potential is expected to soon emerge from the exploitation of new sources of urban data. In the view of many of our interviewees, the issue of ITS is currently primarily a matter of improving many planning and management processes on the basis of digitalized and optimized streams of information, i.e. of 'soft' infrastructures, rather than a matter of long-lasting decisions and investments in hardware. Digitization of maps, zoning attributes, submissions, requests, announcements, etc. often involves a fundamental transformation of related processes, such as in construction licensing procedures, or it allows for completely new processes of coordination, such as the optimization of schedules for construction and maintenance works that fall under the responsibility of different authorities and companies. Such optimization of processes and exploitation of new possibilities apparently requires substantial upfront investment and human capacities, which may be balanced out by efficiency gains only in the long run.

Regarding *styles* "through which the form of the future is disclosed and related to" (Anderson 2010), the ITS Strategy calls only indirectly for *foresight* and *prediction*, when calling for an optimization of the existing "statistical and dynamic database for mobility planning" (Anderson 2010, p. 19). The emphasis there on real-time data, on openness and interoperability of systems, however, indicates that these efforts are motivated more by the prospect of real-time management of traffic flows rather than an optimization of infrastructure investments based on long-term anticipations of mobility demands. And the overall approach of experimentation and staying open to largely unpredictable technological developments can be considered a good example of the style of "*premediation*" which is based in "possibilistic thinking" and a "disclosing of the future as a surprise" (Anderson 2010, p. 782).

As far as the *practices* in which these futures are rendered present and action-able are concerned, remarkably little *calculating* and modelling can be found in the contemporary documents (compared e.g. to the Mobility program from 2013). At the "workshops, ITS fora and ITS exhibitions" (which are mentioned as key means of exchange among the experts in the ITS, Anderson 2010, p. 27), the imagining and visualization of some of the many possible technological futures are presumably a dominant practice. The way in which "small projects" and "experiments" are promoted for testing technologies and forms of cooperation indicates an overall orientation towards learning by *performing*, experiencing and thus embodying possible futures (compare Anderson 2010, pp. 783–787).

Regarding the "*logics* through which anticipatory action is legitimized, guided and enacted" (Anderson 2010, p. 777), we sense an occasional emphasis on *precaution* (e.g. not to invest without testing) but a more pronounced emphasis on *preparedness* (e.g. when the strategy promotes piloting public-private co-operations in order to be prepared for exploiting opportunities that will emerge in the future). Over all, a very optimistic attitude towards the future can be observed given that opportunities are strongly emphasized (e.g. of positioning Hamburg successfully in a competition of places), whereas several risks are identified but are portrayed throughout as being well manageable.

The ITS strategy, and particularly its implementation plan, emphasizes a cooperative approach which brings "economy, research, public authorities and mobility providers" much closer together than before. "Cooperation in public private partnerships will lead to synergy effects" (Anderson2010, p. 27). Sixteen "crucial actors in Hamburg regarding the ITS" are described in detail in appendix 2 of the strategy document (Anderson 2010, pp. 55–60). Public authorities, publically owned utilities and private companies are envisaged to cooperate and likewise "be coopted" in the form of "individual projects, joint support schemes, workshops, ITS fora, ITS exhibitions and as part of a stakeholder process" (Anderson 2010, p. 27). However, the main responsibility for the activities is implicitly delegated to municipal authorities. Additional to the comprehensive "working group on ITS", a high level "steering group" has been established for "assessing the global achievement of objectives, taking major decisions and steering the overall process" (Anderson 2010, p. 28).

A surprising similarity between statements across interviews with members of a multitude of public organizations meanwhile hints at simultaneous processes of change in many of these organizations. The re-shaping of planning processes due to newly available or digitized information seems to contribute to a devaluing of the experience that senior planners accumulate during long professional biographies. Simultaneously, young, flexible and "agile" planning authority staff are in high demand as they possess the will and skills to drive the required interdisciplinary cooperations across institutional boundaries and logics. Their ability to navigate a complex, emerging and fast changing landscape of experimental projects and to quickly establish supportive networks across functional domains and hierarchical levels may be a major driver of their increasing importance compared to long-serving specialists. Within a public administration, which proudly calls itself an "agile, if large corporation", well-connected and flexible mid-career employees seem to be particularly well positioned to determine the future of the transportation infrastructure.

Emphasizing technological and regulatory developments as the main drivers of an unprecedentedly open future can generally legitimize furthermore the (re-) orientation of governance processes towards professional expertise. Promoting an experimental approach can add further legitimacy to an expert-orientation as opposed to transparent governance processes that are open to public scrutiny and participation.

4.1.3 The H2020 Lighthouse Project "mySMARTLife" in Hamburg-Bergedorf

mySMARTLife (mSL) is a five-year project (2016–2021), funded under the European Union's Horizon 2020 research and innovation programme with an overall budget of 20 mio. Euro; thereof about 6 mio. for Hamburg. Twenty-eight partners from 7 countries are collaborating to achieve "sustainable cities with smart people and a smart economy", to reduce CO_2 emissions and to increase the use of renewable energy sources. Demonstration cities are Hamburg, Helsinki and Nantes. In Hamburg, the demonstration site is located in the district of Bergedorf with its 130,000 inhabitants, and the district administration has taken over the coordination of the activities.

Goals in Bergedorf area are to undertake a number of smart city experiments in the fields of housing, energy and mobility. In housing, the experiments comprise e.g. the retrofitting of 500 houses, the offering of a smart home solution and the building of 1,500 new flats. In the field of energy, objectives include the conversion of a district heating system to 100% renewable fuels, the establishment of a smart local heating network with a district cooling component, the establishment of an ice storage, multi-meters and solar thermal, and PV installations. In mobility, measures are smart adaptive lighting for bicycle roads, the installation of humble lampposts, the purchase of ten electrical buses, 40 e-cars and 50 e-bikes for public and private fleets, e-bus charging stations at the bus depot, several fast charging stations and a set of private stations for e-cars all supplied with energy stemming up to 100% from renewable energy sources. Furthermore, it is envisaged to create a multi-modal mobility concept and a car sharing e-community and parcel delivery system in car trunks. Finally, mSL has high ambitions regarding citizen participation, operationalized by extensive options and formats of information and participation in all fields of action, such as e.g. the setting up of an urban platform for dialogue and implementation management, which aims to enable cooperative city development with the needs and expectations of its citizens in mind.

In the terms of Anderson (2010), "the future is disclosed and related to" again in the style of *"premediation"*, which accepts a "disclosing of the future as a surprise" (Anderson 2010, p. 782). The setting up of many experiments in a test-site again reveals a quasi-evolutionary approach. Yet, compared to the ITS framework, we see that mSL comprises a finite list of predefined experiments and an exclusive consortium of beneficiaries regarding the EU funds.

As far as the *practices* through which these futures are rendered present and actionable are concerned, the key idea is obviously—as in the case of ITS—that learning will take place at the individual "test fields". However, mSL additionally aims at learning across the participating cities and beyond. Imagining and visualization are important here, but most important is probably again the idea of learning by *performing*, experiencing and thus embodying possible futures (compare Anderson 2010, pp. 783–787).

Also regarding the *"logics* through which anticipatory action is legitimized, guided and enacted" (Anderson 2010, p. 777), mSL displays features that are very similar to those of the ITS, but dissimilar to the Mobility Program from 2013. Pronounced emphasis is again laid on *preparedness* (e.g. by stressing that learning will enable the exploitation of upcoming opportunities in e-mobility and renewable energy).

The impulse for Hamburg's application to the Smart City call of the EU (H2020-SCC) was given by the Senate Chancellery's office for international cooperation given the mayor's prioritization of the smart city topic. After two earlier applications had achieved very good results but closely missed the threshold for being selected, the city finally succeeded in 2016 in its third attempt. To increase the prospects of the application being successful, the city had hired different consultancies in each stage for managing the consortium. Finally, mSL was accompanied by CARTIF Technology Centre, Spain, which had already successfully coached two earlier Smart City consortia. The Hamburg consortium today includes public partners from the city and the district, partners from science and research, and private companies, e.g. MOIA, a daughter of VW, the public transport provider VHH, Telekom and Hamburg Energy, one of the local energy providers. In addition, the project aims at broad participation of citizens in the research and decision-making process which is organized by a private agency for urban planning and communicative processes at the neighbourhood level.

The activities in the three cities of Hamburg, Helsinki and Nantes include new technological developments in connection with refurbishments of buildings, usage of renewable energies, clean transport and supporting ICT solutions. As one main pillar, the mySMARTLife project aims at reducing the CO_2 emissions of cities, increasing the use of renewable energy sources and making the demonstration cities more environmentally friendly in general. By joining forces, the cities intend to deepen their understanding of technological solutions to common challenges such as smoothening traffic, improving city services and reducing CO_2 emissions, and will use this knowledge to upgrade their infrastructures and deliver new services.

Moreover, the project´s objectives go beyond the environmental dimension and also aim to create more inclusive cities that offer a high quality of life, where citizens play a vital role in the development of the city. In this regard, it is a further objective to develop the concept of "smart people". At the same time, the mySMARTLife partnership also focuses on the "smart economy" approach following the aim to implement an innovative and dynamic economic concept that guarantees employment and adequate income, attracts talents and provides goods and services in line with the actual requirements.

The project started at the end of 2016. Since then, first steps have been taken mainly in the fields of analysis and citizen participation. With regard to the partial electrification of the bus fleet, detailed planning and preparations have been made, including some larger investments in the local bus hub.

Regarding the overall political framework, Hamburg's Smart City agenda has definitely been influenced by the global discourse. The first internal approaches undertaken by the city were focused strongly on modernizing administrative services and processes. In contrast, at the European and international level, Smart Cities were intensively interrelated with the transition to sustainable urban development, e.g. targeting high-end solutions in energy efficiency and renewable energies, e-mobility, waste reduction or efficient use of water. MySmartLife shows how this broad understanding of urban smartness was introduced in Hamburg. Horizon 2020 as a financial incentive motivated the city to apply an integrated approach of ICT innovation and resource efficiency.

However, typical for a Smart City agenda is that it narrows the discussion e.g. about sustainable mobility towards technological and high-tech solutions that correlate with strong economic interests of the automobile industry and ICT companies. Memorandums of Understanding with CISCO in 2014 and Daimler in 2017 underline this classical growth orientation which is affirmed by the strategy and actor constellation in the research project. Both mainly target technological solution finding and related stakeholder networks.

Alternative approaches that could challenge the underlying problem of quantitative growth and resource consumption (e.g. rebound effects) are pushed aside. mySMARTLife seems to strengthen an attitude of ecological modernization rather than transformative social innovation, e.g. by providing proof of concepts for the technological innovations that it tests and promotes. By setting up funding schemes like the Smart Cities call, the EU hence strongly influences the development of urban infrastructures and related discourses.

5   Smart City Futures Shaping Transportation Planning Today

The Smart City ideal prescribes a particular socio-technical future in which various infrastructures are interconnected and fully equipped with ICT, and thus made responsive to an overall optimization on the basis of real-time data from diverse sources. In many European cities, this ideal is now particularly influential in terms of the way mobility infrastructures are governed.

As the above juxtaposition of three contemporary transportation planning frameworks for Hamburg illustrates, those two frameworks that are closely related to smart city discourse share an approach to anticipating the future that is characterized by a style of premediation, practices of performing and a logic of preparedness. This contrasts with the mobility program from 2013, which was developed just before the idea of smart urbanism gained prominence in Hamburg, and which is characterized by the styles of foresight and prediction, practices of calculating and the logic of precaution. From *planning* for a slowly transforming system, the focus has shifted towards the *management* of flows "in real-time" under the condition of ever faster technological change and connected uncertainties (Table 1).

This shift is frequently used as a marker for heralding the innovative strength of Hamburg and for positioning the city as a primary "testfield" for emerging technology, evident e.g. in the application to host the ITS world congress in 2021. Both, the broad attempt of setting up an ITS im Hamburg as well as the international project of mSL are less than two years young. Therefore it is too early to assess their practical outcomes, which will become evident only through the coming months and years, as project years elapse and the ITS world congress in Hamburg comes closer.

Table 1 Comparing three policy frameworks reflecting different attitudes to anticipation

Framework Question	Mobility Program (2014–2020)	ITS-Hamburg (2016–2021)	mySMARTlife (2016–2021)
1) *What approach* to anticipating the future is proposed?	A rational, sequential process from identifying political objectives to implementing related infrastructural change	Development & testing of pilot projects in public private partnerships within a broad framework (test-field due to hosting of ITS world congress 2021)	Selectively implementing predefined experiments on a pilot site (model character) together with European partner cities by members of a triple helix public-private EU H-2020 consortium
a) What *style* is promoted most?	Foresight/prediction	Premediation	Premediation
b) What kind of *practice* is promoted most?	Calculating	Performing	Performing
c) What kind of *logic* is promoted most?	Precaution	Preparedness	Preparedness
2) *Who* is promoted as key actors in preparing for the future?	Local transport authorities embedded in traditional (urban) governance arrangements	An emerging selection of agile specialists (from public & private organizations) who cooperate in emerging pilot projects	A large consortium of public and private partners (as defined in the proposal that was forged in 2015) plus citizens

Compared to the rather conventional mobility program of 2013, however, the two smart city related frameworks clearly highlight different actor groups. The ITS Hamburg with its fast-changing ecology of experiments generally privileges a type of flexible and digitization-oriented employee both within the public administration and in "innovative" companies that collaborate in these projects. And with the project mySMARTlife, a closed consortium was selected to benefit from the "lighthouse" project status and the connected funds, and to consequently shape the implementation of model (mobility) solutions in Hamburg-Bergedorf and beyond. The municipal state government promotes public-private partnerships in order to mobilize both skills and capital from the private sectors. The usage of the term "innovation eco-system" as an analogy for biological evolution reflects this approach in mySMARTlife, as in many Smart City initiatives worldwide.

At the same time, the administration of Hamburg claims for itself a final say in all major decisions to ensure that the public interest is sufficiently considered (cf. speeches of Hamburg's Lord Mayor Olaf Scholz 2014, 2016).

Whereas in the recent past there was an incumbent system of planning on the basis of more or less agreed upon future scenarios, which could at least theoretically be opened up for the involvement of citizens, the experiments that are currently conducted under the flag of proactive and experimental digitization are necessarily accessible only to a few governmental and commercial actors. Portraying the mobility future as being shaped by fast changing technologies and by exploding expectations regarding inter-modal real-time information and algorithm-based anticipation hence shapes not only the contemporary smart city experiments but simultaneously de-legitimizes the conventional and potentially participatory approaches to planning on the basis of politically agreed upon objectives.

The EU money channelled through Smart City calls in the Horizon 2020 framework program seems to deliberatively foster such re-framings of infrastructure planning and the respective roles of urban governments and other actors. These funds provide primary opportunities for both local governments and private actors to gain more influence in these processes (e.g. vis a vis parliaments and citizens).

Whether EU money is involved or not, the focusing of financial and human resources on projects that fit with the "smart city" or "proactive digitization and innovation strategy" may, under the given conditions of scarce resources, very likely result in a comparative weakening of low-tech or alternative ambitions, and in a down-playing of questions of sustainability and equal access to mobility.

In recent years, the hype surrounding smart innovation has prompted municipalities to keep abreast of new developments and to future proof their administration and infrastructures. There seems to be an associated narrowing down of solutions and problem solving corridors. Driven by business actors and their respective interests, high-tech solutions moved into the core of the discussion and often seem to be without alternative. This hegemonic dominance can be found in the two most recent frameworks discussed above where Hamburg has followed the mainstream of technological innovation. According to our observations, the EU and its funding schemes can play a strong role in mainstreaming a particularly technology-oriented development path.

6 Conclusions

This book seeks to develop an understanding of ways in which particular visions of socio-technical futures can influence the contemporary shaping of socio-technical systems. In this chapter, we provide an illustrative example of such a process. Smart City futures commonly feature bright prospects for multi-modal mobility on the basis of interconnected, driverless vehicles, and public means of transportation that are optimized based on real-time information. As many of the technologies involved in these scenarios, and particularly the internet-of-things, involve high risks for security and privacy, acceptable pathways towards this future are not easy to determine. In consequence, smart mobility initiatives often follow an experimental approach, and promote the development of multiple experiments including new public-private co-operations in order for all participating actors to be prepared to exploit the opportunities that technological developments will bring.

As a flipside of the emphasis that smart mobility initiatives commonly place on collaborative experimentation, other views on how an appropriate development of transportation infrastructure may be determined are devalued. We detected such a shift in the context of Hamburg by contrasting changing attitudes towards anticipating urban mobility futures in three contemporary frameworks of transportation planning. With smart urbanism gaining prominence in Hamburg, an attitude of planning characterized by the styles of foresight and prediction, by practices of calculating and the logic of precaution was replaced or at least complemented and challenged by an attitude of experimenting towards real-time management that is characterized by a style of premediation, practices of performing and a logic of preparedness.

We discussed multiple implications of such a shift on governance arrangements and prospects for democratic participation and decision making. From our perspective, it is questionable that a purely expert driven process will be the most effective pathway towards robust decision making in such sensitive issues as future opportunities for mobility and privacy in public spaces. Hence, maybe the candidate 'solutions' for future mobility demands should be exposed to societal concerns at an earlier stage in their development. The literature on (constructive) technology assessment contains ample debates on how such an 'upstream' engagement could be achieved through public participation and dialogue, and what typical challenges, dilemmas and limitations of such an engagement are.

What makes the topic of data-driven mobility management particularly contested is the fact that the more traffic management will be conducted on the basis of (hidden) algorithms and real-time data (of often questionable quality), the more difficult it will become to contest the orientation and priorities of this management in a democratic process. This in turn makes it unlikely that the future development of urban infrastructures well reflect societal preferences.

Regarding the analytical challenges that this book starts to address, we conclude that it proved very fruitful to comparatively scrutinize programs for transportation planning—understood as results of contested institutionalization processes –, for the emphasis that they each place on a particular set of styles, practices and logics in the attitude that they prescribe towards the anticipation of future developments and demands.

It was furthermore insightful to pay attention to what actors were highlighted in the context of these programs, and which groups gained and which lost influence after smart city related visions of future mobility gained influence in Hamburg. A challenge seems to be that the engagement of commercial actors—who are naturally interested in the development of markets for their solutions and hence place a strong emphasis on opportunities rather than challenges—holds the danger that necessary public debates about objectives and acceptable and unacceptable risks are prevented by public-private alliances or largely biased in favour of high-tech solutions.

Thus, it seems very worthwhile to explore more deeply the in- and exclusiveness of the actor constellations in such processes, e.g. regarding the role of coincidences or strategies behind the selection of partners, including a possible gate-keeper function of the process initiator. Of relevance are also a dynamic perspective and the consideration of scope for reflexivity, as well as impacts on policy making and technology transfer.

To problematize such processes, and maybe even to intervene in favour of alternative, more open, inclusive and productive governance arrangements, seems to be what STS informed scholars can and should contribute to safeguarding democratic control of technology development now and in the future.

References

Albino, V., Berardi, U., & Dangelico, R. M. (2015). Smart cities: Definitions, dimensions, performance, and initiatives. *Journal of Urban Technology, 22*(1), 3–21.

Anderson, B. (2010). Preemption, precaution, preparedness: Anticipatory action and future geographies. *Progress in Human Geography, 34*(6), 777–798.

Berliner Erklärung (o. A.). (2017). *Berliner Erklärung zu Forschung und Innovation für eine nachhaltige urbane Mobilität—Neues wagen! Mehr Mut für innovative Wege in der Mobilität.* Berlin: BMBF.

Brown, N., Rappert, B., & Webster, A. (2000). Introducing contested futures: From "looking into" the future, to "looking at" the future. In N. Brown, B. Rappert, & A. Webster (Eds.), *Contested futures—A sociology of prospective techno-science* (pp. 3–20). Burlington: Ashgate.

Bulkeley, H., & Castán Broto, V. (2013). Government by experiment? Global cities and the governing of climate change. *Transactions of the Institute of British Geographers, 38*(3), 361–375.

Bulkeley, H., Coenen, L., Frantzeskaki, N., Hartmann, C., Kronsell, A., Mai, L., et al. (2016). Urban living labs: Governing urban sustainability transitions. *Current Opinion in Environmental Sustainability, 22,* 13–17.

BWVI Hamburg. (2016). Verkehr 4.0 – ITS-Strategie für Hamburg – Strategie zur Weiterentwicklung und Umsetzung von Maßnahmen Intelligenter Transportsysteme (ITS) in Hamburg. https://www.google.de/url?sa=t&rct=j&q=&esrc=s&source=web&cd=1&cad=rja&uact=8&ved=0ahUKEwj67cLe37_SAhVEOxQKHXKdAroQFggpMAA&url=http%3A%2F%2Fwww.its2021hamburg%2Fdownloads%2FITS%2520Strategie%2520Hamburg.pdf&usg=AFQjCNFv3hc66Cs64GKviikO97p81DJS6A&sig2=D_sJyEqyB_Mslt4cFQ_Nfg.

Caragliu, A., Del Bo, C., & Nijkamp, P. (2011). Smart cities in Europe. *Journal of Urban Technology, 18*(2), 65–82.

de Jong, M., Joss, S., Schraven, D., Zhan, C., & Weijnen, M. (2015). Sustainable–smart–resilient–low carbon–eco–knowledge cities; making sense of a multitude of concepts promoting sustainable urbanization. *Journal of Cleaner Production, 109,* 25–38.

Evans, J., Karvonen, A., & Raven, R. (2016). *The experimental city.* London: Routledge.

Goldsmith, S., & Crawford, S. (2014). *The responsive city: Engaging communities through data-smart governance.* John Wiley & Sons, ISBN: 1118910931, 9781118910931.

Grunwald, A. (2011). Energy futures: Diversity and the need for assessment. *Futures, 43*(8), 820–830.

Grunwald, A. (2012). *Technikzukünfte als Medium von Zukunftsdebatten und Technikgestaltung* (Vol. 6). Karlsruhe : Karlsruher Studien Technik und Kultur.

Hamburger Senat. (2013). *Mobilitätsprogramm 2013 – Grundlage für eine kontinuierliche Verkehrsentwicklungsplanung in Hamburg.* Hamburg: Senat der Stadt Hamburg. http://www.hamburg.de/bwvi/mobilitaetsprogramm/.

Jasanoff, S. (2015). Future imperfect: Science technology, and the imaginations of modernity. In S. Jasanoff & S.-H. Kim (Eds.), *Dreamscapes of modernity: Sociotechnical imaginaries and the fabrication of power* (pp. 1–33). Chicago: University of Chicago Press.

Kitchin, R. (2014). The real-time city? Big data and smart urbanism. *GeoJournal, 79*(1), 1–14.

Kourtit, K., Nijkamp, P., & Partridge, M. D. (2012). The new urban world. *European Planning Studies, 21*(3), 285–290.

Meijer, A. (2017). Datapolis: A public governance perspective on "Smart Cities". *Perspectives on Public Management and Governance, 1*(3), 195–206.

Mora, L., Bolici, R., & Deakin, M. (2017). The first two decades of smart-city research: A bibliometric analysis. *Journal of Urban Technology, 24*(1), 1–25.

Raven, P. G. (2017). Telling tomorrows: Science fiction as an energy futures research tool. *Energy Research & Social Science, 31,* 164–169.

Raven, R., Sengers, F., Spaeth, P., Xie, L., Cheshmehzangi, A., & de Jong, M. (2017). Urban experimentation and institutional arrangements. *European Planning Studies, 24,* 258–281.

Scholz, O. (2014). "Smart City Initiative/Memorandum of Understanding". Hamburg, FHH. http://www.hamburg.de/contentblob/4306512/eec2665f63e2b922b249769923713987/data/2014-04-30-smart-city.pdf. Accessed 30 April 2014.

Scholz, O. (2016). Universitätsgesellschaft – Digitale Stadt Hamburg. Hamburg. http://www.hamburg.de/buergermeisterreden-2016/5965618/2016-05-02-universtaetsgesellschaft/. Accessed 2 May 2016.

Schwanen, T. (2017). Geographies of transport II. *Progress in Human Geography, 41*(3), 355–364.

Sengers, F., Berkhout, F., Wieczorek, A., & Raven, R. (2016). Experimenting the city: Unpacking notions of experimentation for sustainability. In J. Evans, A. Karvonen, & R. Raven (Eds.), *The experimental city*. London: Routledge.

Sengers, F., Späth, P., & Raven, R. (2018). Experimenting with smart eco-cities in Dutch and German cities: Discourses, institutions, materiality. In S. Marvin, H. Bulkeley, Q. L. Mai, & K. Mccormick (Eds.), *Urban living labs: Experimentation and socio-technical transitions* (pp. 74–88). New York: Routledge.

Sennett, R. (2012). No one likes a city that's too smart. The Guardian. http://www.guardian.co.uk/commentisfree/2012/dec/04/smart-city-rio-songdo-masdar.

Shelton, T., Zook, M., & Wiig, A. (2015). The 'Actually existing Smart City'. *Cambridge Journal of Regions, Economy and Society, 8,* 13–25.

Song, C. H., Elvers, D., & Leker, J. (2017). Anticipation of converging technology areas— A refined approach for the identification of attractive fields of innovation. *Technological Forecasting and Social Change, 116,* 98–115.

Spaeth, P., Hawxwell, T., John, R., Li, S., Löffler, E., Riener, V., et al. (2017). *Smart eco-cities in Germany: Trends and city profiles (Smart-eco project)*. Exeter: University of Exeter.

Townsend, A. M. (2013). *Smart cities: Big data, civic hackers, and the quest for a new utopia*. New York: Norton.

van Lente, H. (1993). *Promising technology: The dynamics of expectations in technological developments—PhD thesis*. Enschede: University of Twente.

Vanolo, A. (2014). Smartmentality: The smart city as disciplinary strategy. *Urban Studies, 51*(5), 883–898.

Vanolo, A. (2015). Smart city and urban development: Note for a critical agenda. *Scienze del Territorio, 3,* 111–118.

Vanolo, A. (2016). Is there anybody out there? The place and role of citizens in tomorrow's smart cities. *Futures, 82*(Supplement C), 26–36.

White, J. M. (2016). Anticipatory logics of the smart city's global imaginary. *Urban Geography, 37*(4), 572–589.

Zook, M. (2017). Crowd-sourcing the smart city: Using big geosocial media metrics in urban governance. *Big Data & Society, 4*(1), 2053951717694384.

Philipp Späth (Ph.D. and habilitation) is geographer, political scientist and STS scholar and senior researcher and head of the research group "urban environmental governance" at the Institute of Environmental Social Sciences and Geography of Albert-Ludwigs-University Freiburg.

Jörg Knieling (Ph.D.) is spatial planner and political scientist, professor and head of the Institute of Urban Planning and Regional Development and dean of the urban planning program at HafenCity University Hamburg.

Part III
Intervening into the Present Through Prospective Reasoning

Reflexive Hermeneutics Against Closing Down Technology Assessment Discourses: The Case of Synthetic Biology

Go Yoshizawa

Abstract

Policymakers and scholars have employed technology assessment to formulate and deliberate policy and social agendas by anticipating plausible futures for the last half century. Proposing a reflexive hermeneutics approach and taking a case of synthetic biology, this study illustrates how TA discourses are socio-culturally and institutionally bounded. The case study leads to the conclusion that future-oriented conceptions of the 'precautionary' and 'proactionary' intertwine with general concepts including biosafety, biosecurity, bioeconomy, and biodiversity. These general concepts together with the advent of disruptive innovation in synthetic biology have already affected perception of possible futures.

1   Introduction

Policymakers and scholars have employed technology assessment (TA) to formulate and deliberate policy and social agendas by anticipating plausible futures for the last half century. Aiming to support policy and social decision-making, the selection of technology and the direction of futures studied in this activity are, however, always at risk of being affected by incumbent interests and needs.

G. Yoshizawa (✉)
Division of AFI, Oslo Metropolitan University, Oslo, Norway
e-mail: Go.Yoshizawa@oslomet.no

© Springer Fachmedien Wiesbaden GmbH, part of Springer Nature 2019   189
A. Lösch et al. (eds.), *Socio-Technical Futures Shaping the Present*,
Technikzukünfte, Wissenschaft und Gesellschaft / Futures of Technology,
Science and Society, https://doi.org/10.1007/978-3-658-27155-8_9

Where TA often takes a discourse-analytic approach, TA discourses themselves have been rarely studied. How are TA discourses politicized, and in what socio-cultural and institutional contexts? This research question is a serious one given that most dedicated TA institutions are still too often regarded as a 'neutral' observer, positioned at a safe distance from controversy. This study aims to reveal that even the socio-technical futures depicted in TA studies are not free from present social and political concerns. The introductory section of this chapter gives a brief overview of how futures and politics are incorporated or embedded in the development of TA. After introducing the 'hermeneutic turn' in perspectives on the future and reviewing how various hermeneutic approaches have been disseminated in the social sciences, it then proposes a 'reflexive hermeneutics', a new conceptual and methodological framework. The second chapter takes as a case study TA on synthetic biology, that is, the design and construction of new biological parts, devices, and systems or the redesign of existing natural biological systems. Synthetic biology is a key emerging technology field, in which various experts and stakeholders have put forward and disseminated far-ranging visions. The selection of synthetic biology as a case is not only due for practical reasons, as there is only a relatively short history of scientific progress, policy development, and societal debates around it but also because, at the same time, there is a wide variety of political and societal discourses found in this field. Synthetic biology has yielded many future visions, promises, hopes, and fears but nevertheless remains in the early phase of development and application. The case study leads to the conclusion that future-oriented conceptions of the 'precautionary' and 'proactionary' intertwine with general concepts including biosafety, biosecurity, bioeconomy, and biodiversity. These general concepts together with the advent of disruptive innovation in synthetic biology have already affected perception of possible futures.

1.1   Futures and Politics in Technology Assessment

Attempts to construct socio-technical futures in technology assessment (TA) can be traced back to early efforts in the 1960s to forecast technological developments. However, another approach, which has been called TA, promotes awareness of the potential negative or undesirable social and economic consequences of technological development. It analyses societal effects of technological change and provides an understanding of the possibilities for social choice relating to that change (Cronberg 1996). These two approaches are retrospectively regarded as 'traditional TA' characterised by a deterministic vision and prognostic imagination of future technology. This prognostic approach might be observed to be the primary motivation behind the establishment of the Office of Technology Assessment (OTA) in the United States in 1972, which aimed to provide Congressional members and committees with scientific advice. However, the traditional TA concept started to lose ground in the mid-1970s, when two major problems arose. First, it became clear in practice, as well as from theoretical considerations, that the impacts of technology can only very partially be foreseen and also that well-developed technology is difficult to direct because it has become 'entrenched' in society (Collingridge 1980). Second, traditional TA did not provide decision makers with balanced, complete, or objective information, as in the case of the recombinant DNA or nuclear-energy controversies in the United States. TA activity conceived in this way is essentially an interface between science and policy and thus faces all the problems of science as well as those of policy (van Eijndhoven 1997).

This boundary work between science and policy thus entailed a new, more broad-based and process-oriented TA in the late 1970s, the framing and targeting of which were extended to include participation by actors from outside the scientific community. TA has since become progressively more tuned to decision-making processes, and has come to involve more intensive interaction between practitioners and decision-makers. This new form of TA plays an active, strategic role in the development of technologies and their applications, like a tracker dog, whereas the traditional, reactive, early warning TA sought to provide a 'neutral' or 'objective' input in decision-making, like a watchdog (e.g. van den Ende et al. 1998). Further, since the late 1990s, a specific institutionalised mode of this new TA has emerged, referred to as parliamentary TA, which focuses on TA institutions under parliamentary bodies, in particular in EU member states (Vig and Paschen 2000). Given the priority of autonomy over relevance, the institutionalisation of TA in Europe has led to significant changes in the way issues are framed, methods designed, and participants selected in the assessment processes. These changes represent a shift in the political handling of the relationship between science and society.

1.2   Hermeneutic Turn

The process-oriented form of TA described above has since developed into a scenario-based deliberative process for identifying potential impacts of emerging technology. As new technological developments may take unexpected turns, necessitate structural adjustments (Freeman and Perez 1988), and/or lead to unintended consequences, scenario development risks falling into mere articulation of wishful thinking. By opening public discussions about possible and desirable directions of emerging technologies, TA studies can then be considered to reveal our present perspectives on the future and by doing so also to contribute to shaping the present. The claim of a radical 'immanence of the present' (Grunwald 2014a) casts lively and controversial debates about technology not as anticipatory, prophetic, or quasi-prognostic talk of the future, but as an expression of our present-day situation; that is, the subject of investigation is not what is being said, with whatever indeterminate degree of justification, about coming decades, but what is revealed about us by the fact that these debates are happening today. This change of perspective has been called a 'hermeneutic turn' (Grunwald 2014b); it renders visible and amenable to critical discussion the preconditions and the available degrees of freedom for alternative trajectories of technological development where these preconditions do not refer to 'what will the future bring' or 'what might one day be the case' but to today's expectations, problem formulations, wishes, and values, which (are taken to) literally inform technological developments. This observation then also raises the question of the functions of TA in current debates and discourses. To trace the possibilities of such a hermeneutic orientation, it is necessary to reveal the reasons for and sources of divergence of projected futures. A hermeneutic orientation is about trying to learn something about ourselves—our societal practices, subliminal concerns, implicit hopes and fears—from the diversity and divergence of future(s) studies.

Hermeneutic orientation can have two different modes: the former projecting a bounded divergence of futures (mode [a]), and the other oriented to futures that are completely open, unbounded, and opaque (mode [b]) (Lingner 2017). Inspired by Luhmann (1982), Adam and Groves (2007) distinguish 'present futures' from 'future presents'. The present future refers to perspectives of the future from the standpoint of the present, aiming to predict, transform, and control the future for benefit of the present; in contrast, the future present is a conceptualised future, which sees the future already unfolding in the present in line with our own activities and closes down options for present and successor generations. When the gradation from the prognostic mode to hermeneutic mode [b] appears to become more supportive of mentally represented utopias or dystopias that open up present futures, envisioning the future inevitably remains path-dependent and bounded in hermeneutic mode [a]. This kind of discussion is cognate with 'opening up' the outputs of TA through plural and conditional advice (Ely et al. 2014). What is critical to understand is that the present discourse of the future is very likely to be affected by closing down pressures from intended and unintended processes and power mechanisms that have been discussed in many ways in the study of society, history, philosophy, politics, and economics (Stirling 2008). More participatory and inclusive science policy thus paradoxically leads to a 'subtle but powerfully less obvious, pre-emptive closing down' (Wynne 2016, p. 103), which has been called the 'hermeneutic imperialism' of science (Wynne 2014, p. 62). The scope and integrity of TA are constrained by implicit political commitments and interests (Wynne 1975); they can also be driven by public controversies in an informal manner (Rip 1986), which may push the public to self-organise and issue 'popular TA' when official processes fail to deliver the level of accountability they desire (Jasanoff 2003).

TA can serve as a fruitful focus of this kind of study in that they are supposed to take into account divergent perspectives on uncertain techno-visionary futures as much as possible. Understanding the limitations of TA studies in illustrating these divergent futures may hint at a resolution to the 'conventionalization' of our discourse on the future and governance in the present. Narratives taking the form of visions, expectations, hypes, hopes, fears, or dystopias appear to direct us to radicalized futures, either better than the best case or worse than the worst case described in conventional scenario analysis. In this light, the present study serves as a test case on if and to what extent the hermeneutic approach can be empirically distinguished from prognostic or scenario-based approaches.

1.3   Reflexive Hermeneutics

Few would doubt that certain similarities and links exist between TA studies and policy analysis, for both entail policy-oriented analysis of institutions (Majone 1977). In policy analysis, the hermeneutic approach is not new. Traditionally, as is well known, analytic tools are often treated as 'value free', and facts are held to be separable from values; this kind of positivist foundation for policy analysis has received numerous and varied criticisms (e.g. Rein and White 1977; Torgerson 1986). Some critics instead propose hermeneutic approaches to policy analysis (Fischer 2003); these are problem-oriented, contextualized, multidisciplinary in approach, and explicitly normative in perspective (deLeon 1988, p. 7), as well as more participatory, deliberative, and democratic (Durning 1999).

Critical hermeneutics is one of the main approaches to hermeneutic policy analysis (Dryzek 1982; Fischer 1998). It emphasizes the interpretive, conditional character of knowledge and its goal is to contribute to a transformational and emancipatory conception of social science (Morçöl 2002). However, this kind of critical approach tends to offer only reductionist proof of the presence of ideological notions and functions in concrete language (Keller 2005). As a remedy, two other hermeneutic approaches to the social science have been developed, mostly in German-speaking countries—objective hermeneutics and sociological hermeneutics. Objective hermeneutics focuses cognitive interest on action-generating latent meaning structures (Hitzler 2005; Reichertz 2004). It assumes that there is a strong and mutually responsive connection between the conceptual and actionable realms. From a practical point of view, rendering objective structures of everyday interactions by revealing the latent meanings of a text is very time- and labour-intensive (Wodak 2011). In contrast to critical hermeneutics, sociological hermeneutics, or hermeneutic sociology of knowledge, follows the socio-phenomenological research tradition based on the works of Schütz, Berger, and Luckmann and reflects the criticism of the 'metaphysics of structures' in objective hermeneutics. As a new line of discourse analysis crucially informed by Foucault's pioneering work (Angermüller 2011), it aims to discover the inter-subjective meaning of actions, (partly) equated with the (anticipated) willingness to react set up by the action within an interaction community (Reichertz 2004).

Reflexive hermeneutics is less established and recognized than the other schools, embracing wider academic origins and notions in sociology, psychology, and ethnography. It moves beyond the inherited opposition between objective hermeneutics, which focuses on uncovering the intended meaning, and critical hermeneutics, which focuses on uncovering the true meaning behind the apparent, distorted understandings of society. Where antecedent literature is likely to concern individual introspection (in epistemological terms) and intersubjective interaction (in methodological terms), this study takes reflexive hermeneutics rather as a systematic process of exploration of the prior commitments framing knowledge (Wynne 1993). Similar to sociological hermeneutics, reflexive hermeneutics accounts for intersubjective social narratives developed through collective identity discourse and embodiment (Cunliffe 2011) and gives expression to social imaginaries in which language and social practices are embedded. It is *reflexive* in the sense that social imaginaries become a subject of moral and political reflection and dialogue (Gibbons 2008).

2   Case: Synthetic Biology

2.1   Method

The case study conducts discourse analysis of documents on TA studies on synthetic biology. Among a diverse range of publications on synthetic biology, it selects 35 documents published in 2005–2015 (see Document List) according to the following criteria:

1. They are written in English, as this language is the most influential in the globalized policy arena and also provides a reasonable practical limit to the potential data.
2. They discuss a variety of implications and aspects of synthetic biology (e.g. social, economic, legal, political, ethical, cultural), and do not just focus on scientific and technological aspects.
3. They address decision-makers and policymakers, or they appear to have substantive impact on policymaking contextually.

In order to relativize and evaluate the role of dedicated TA institutions (here, parliamentary TA organisations and networks), the analysis also includes policy analyses of similar kinds provided by other institutions. Accordingly, the selected 35 documents show a wide variety in terms of focus and affiliation.

This analysis first takes a grounded approach to these documents by means of qualitative data analysis (MAXQDA12), through which paragraphs, phrases, and words are manually coded. The coding does not cover the whole documents but focuses on discourses on socio-technical futures (cf. Grunwald 2014b). Reflecting the sociology of knowledge approach to discourse (Keller 2005, 2011, 2012), discourse analysis is not just textual analysis but 'case study, observation, and even a dense ethnographic description' linking the social and institutional dimensions of knowledge production and circulation. It complements intra- and inter-document analysis with reference to earlier academic studies in order to examine the social, political, and historical context of the discourse (cf. Li et al. 2015). The reflexive hermeneutic approach systematically explores and compares two specific socio-cultural and institutional settings that seem to differently frame TA discourse— namely, transatlantic perspectives and institutional TA initiatives.

A simple content analysis that search for future-looking or temporal terms like 'future', 'vision', 'hope', 'expectation', 'hype cycle', 'precautionary', 'proaction-ary', 'preparedness', 'proactive', 'prospective', and 'monitoring', and related key terms like 'proportionality', 'prudent vigilance', 'dual use', 'biosecurity', 'bioeconomy', 'biodiversity', 'playing God', 'creating life', 'naturalness', and 'sustainability', is combined with the examination of a tentatively set code and of the usage of concepts, definitions, metaphors, and rhetorical figures, in reference to future implications, aspects, risks and prospects.

2.2   Analysis

Reflecting the result of the qualitative data analysis, this section distils divergent discourses on steering future development of synthetic biology. The first sub-section (2.2.1) critically examines conventional understanding of transatlantic differences on governance for the future. The following two subsections (2.2.2 and 2.2.3) then discuss this perceived contrast as rooted in general concepts of synthetic biology, including biosafety, biosecurity, bioeconomy, and biodiversity. The final two subsections (2.2.4 and 2.2.5) focus on the hype cycle as a projected future by illuminating how TA institutions have weaved hype and exaggerated claims into their discourses.

2.2.1 Precautionary-Proactionary Debate

Precaution cannot be seen as a simple dividing line for transatlantic discord; for instance, while the US Environmental Protection Agency is especially precautionary on nuclear waste deposits, the EU has undertaken highly precautionary policies on GM foods (Wiener et al. 2011). Historically, EU laws on GMOs were clearly based on the precautionary approach of 'case-by-case' and 'step-by-step' that had been advocated by the OECD (OECD 1986, p. 42; Cantley and Lex 2011, p. 48; von Schomberg 2006, p. 26).

When van Est et al. proposed a paradigm shift from GM to synthetic biology, they assumed that ethical and social implications would be inherited over the shift (van Est et al. 2007, p. 2), and in fact, precautionary approaches to synthetic biology are framed in a similar vein to those to GM in many European policy documents (de Vriend 2006, p. 53; COGEM 2008, p. 4; EGE 2009, p. 40, 65, 77; COGEM 2013, p. 53). Attributing the origin of the precautionary principle to European contexts, the US Presidential Commission for the Study of Bioethical Issues (PCSBI) introduced the 'proactionary principle' as a contrasting perspective. The basic idea of this principle is that 'emerging science and technology should be considered safe, economically desirable and intrinsically good unless and until it is shown to be otherwise' (Parens et al. 2009, p. 18). PCSBI tactically contrast precautionary Europe with proactionary America in popular discourse and then propose 'a middle ground—an ongoing system of *prudent vigilance* that carefully monitors, identifies, and mitigates potential and realized harms over time' (PCSBI 2010, p. 124). The notion of 'prudent vigilance' may relate to the idea of 'preparedness' (PCSBI 2010, p. 128); (Keasling 2010), which, contrasting with 'safety' (which addresses dangers through safeguards and procedures) and 'security' (concern about challenges related to the political environment), requires vigilant observation, regular forward thinking, and ongoing adaptation in the face of uncertainty (Rabinow et al. 2006). Although the conventional contrast between precautionary EU and regulation-averse US is also found in the OECD report (OECD 2014, p. 126), the proactionary principle seems to have been an artificial counterpart of the precautionary principle in the post-9/11 US. The precautionary-proactionary dichotomy can overlap with the security-freedom dichotomy, and prudent vigilance is resurrected as a golden mean to partially satisfy the need highlighted by the logic of preparedness.

Both prudent vigilance and precaution deal with uncertainty (Kaebnick et al. 2014); as Science and Technology Options Assessment (STOA) observe, these are 'intended to raise awareness to as yet unknown developments' (van Est and Stemerding 2012, p. 207), but the approaches involved are different. Like the American interpretation of 'safety', emphasizing prevention and protection, the precautionary approach focuses on pre-incident measures. For instance, within Europe, contemporary strategies for managing biorisk are various, termed for instance contingency planning (France), protection (Germany), or resilience (UK) (Lentzos and Rose 2009), but they all appear to depend greatly on their ability to protect or handle any disruptive challenges, and to plan necessary measures beforehand. On the other hand, the prudent vigilance framework eyes post-incident activities, assuming that unexpected consequences are inevitable. Unlike GMOs, which are normatively defined as *a priori* potentially hazardous in a precautionary regulatory framework (von Schomberg 2006, p. 26), synthetic biology often raises issues around the trade-off between risk and return (COGEM 2008, p. 18) or between risks (EGE 2009, p. 65; CBE/CNECV, p. 18).

The trade-off debate calls for replacement of the precautionary principle with the *proportionality principle*: that 'research methods are necessary to the aims pursued and that no alternative more acceptable methods are available' (EGE 2000, 2003); thus, the proportionality principle addresses the relation between ends and means (Hermerén 2012, p. 374). Rath (2014) insists, given that a ranking of incommensurable values of security and fundamental rights is not foreseeable, the ability of the precautionary principle to resolve such dilemmas is limited. Thus, for him, restricting the publication of the gain-of-function experiments on H5N1 influenza, for instance, is an un-proportional response. The Netherlands Commission on Genetic Modification (COGEM) also refers to the proportionality principle as 'the balance between the risks and benefits to society' (COGEM 2013, p. 49). Some EU documents discuss the precautionary principle in the context of risk management (de Vriend 2006, p. 53; CBE/CNECV 2011, p. 28; COGEM 2013, p. 53) or impact assessment (EGE 2009, p. 77), while others insist that the estimation of non-calculable risks should be counterweighed by that of potential benefits from 'scientific and commercial progress' (COGEM

2008, p. 18) so as not to 'seriously affect innovation' (BBSRC 2010, p. 43). Similarly, Rerimassie and Stemerding address this issue as a matter of weighting in the assessment of biosafety and biosecurity risks of synthetic biology (Rerimassie and Stemerding 2014, p. 75). However, the Nuffield Council on Bioethics (NCB) criticises such discourse for inviting inappropriate application of the precautionary principle to the single dimension of risk management (NCB 2012, p. xxv–xxvi).

2.2.2 Dual Use and Biosecurity

The difficulty of balancing risks and benefits to society is often expressed through the term 'dual use', which dates back at least to 1993, in a report by the US OTA on weapons of mass destruction and the technologies underlying them (Forge 2010). van Est et al. touch on dual use research in relation to the establishment of the National Science Advisory Board for Biosecurity (NSABB) in 2004 (van Est et al. 2007), triggered by a recommendation in the report 'Biotechnology Research in an Age of Terrorism' (Committee on Research Standards and Practices to Prevent the Destructive Application of Biotechnology 2004). This so-called 'Fink report' introduces the term 'dual use dilemma', 'in which the same technologies can be used legitimately for human betterment and misused for bioterrorism' (p. 1). In the US context, this dilemma is regarded as social and political in the face of the perceived threat of bioterrorism, due to which 'synthetic biology should be subject to institutional review and oversight' (NSABB 2010, p. iii). In Europe, by contrast, it tends to be interpreted as ethical (EGE 2009, p. 80; SYNTH-ETHICS 2011, p. 6); (Kuhlau et al. 2008), requiring education for individual professionals (EASAC 2010, p. 19) and self-regulation (de Vriend 2006, p. 48, 55).

This contrast probably comes from a different focus of concern—the US focuses strongly on biosecurity whereas the emphasis in Europe is on biosafety, ethics, and public engagement, as a direct consequence of the GM-food debate (e.g. Pauwels and Ifrim 2008). Some European policy documents admit relatively low concern for biosecurity in Europe (de Vriend 2006, p. 54; Bunn 2008, p. 3). Most arguments are actually borrowed from the US context, and follow a dominant frame of discourse that assumes that the potentially disruptive character of biotechnology increases the threat (EGE 2009, p. 67; OECD/RS 2010, p. 30). This kind of framing is premised on a linear model of technology transfer (McLeish and Nightingale 2007); however, bioweaponisation is not a linear process that moves smoothly from one stage to the next (Marris et al. 2014). The poliovirus and phiX synthesis experiments were based on tacit knowledge, laboratory practices, team-work, and trouble-shooting efforts rather than on any completely new and unforeseen technological capability (Vogel 2008). PCSBI's report already points out such contingencies and socio-technical factors in order not to foment biosecurity threats (PCSBI 2010, p. 72).

2.2.3 Bioeconomy and Biodiversity

In the US, bioeconomy is defined as 'all around us', producing new drugs and diagnostics for improved human health, higher-yielding food crops, emerging biofuels to reduce dependency on oil, and biobased chemical intermediates (White House 2012, p. 7). In Europe, in contrast, bioeconomy is vaguely defined as a key element for smart and green growth (European Commission 2012), to which synthetic biology is nevertheless expected to contribute (UK Synthetic Biology Roadmap Coordination Group 2012, p. 27; van Est and Stemerding 2012, p. 164). However, looking back at the GM debate, the OECD warn that growing support in Europe for the idea of a future bioeconomy built on the technological platform of synthetic biology creates a quandary if public opinion rejects synthetic biology (OECD 2014, p. 28). Curiously, in fact, except for the UK *Roadmap*, European policy documents on synthetic biology almost completely ignore the idea of bioeconomy. This disregard may be cognate with the low-key public debate on biodiversity related to synthetic biology (Sauter 2011, p. 20–21), in the sense that both biodiversity and bioeconomy concern security and sustainability of global natural and innovation ecosystems (Sheppard et al. 2011).

Given explicit and implicit efforts to avoid a repeat of the GM debate and the consequential policy of putting biosafety before biosecurity European discourses on biodiversity adopt a low-key, cautious attitude. For instance, the Biotechnology and Biological Sciences Research Council (BBSRC) state that 'it was acknowledged that uncertainties still remain regarding the impact of GMOs on biodiversity and the likelihood of gene transfer' (BBSRC 2010, p. 78), while COGEM admit the need of research on biodiversity caused by the release of synthetic organisms into the environment (COGEM 2013, p. 52). Where synthetic technologies on ecology are double-edged, negative implications are often emphasised; for instance, the European Group on Ethics in Science and New Technologies (EGE) insist that '[s]ynthetic biology products must ... address biosafety issues when they have consequences for ecology and human health' (EGE 2009, p. 64). Also, in the Spanish Bioethics Committee (CBE) and Portuguese National Ethics Council for the Life Sciences (CNECV)'s report, the possibility of creating a hybrid or chimerical is viewed as ideally not 'a question of preserving biodiversity, but rather of limiting it, insofar as synthetic biology could influence it, with the aim of preserving the identity of species' (CBE/CNECV 2011,

p. 13). Nonetheless, irreversible risks to the environment, characterised as a key element of the precautionary principle, are never highlighted in European policy documents in the context of synthetic biology (see [EGE 2009, p. 65; CBE/ CNECV 2011, p. 17; Rerimassie and Stemerding 2014, p. 33] in the context of GM).

By contrast, US discourses address a wide range of biodiversity issues in terms of both prospects and concerns, reflecting high market expectations and security concerns about synthetic biology. On the sunny side, the exploration of biodiversity is said to need more extensive environmental sampling (NAKFI 2010, p. 58) and to lead to the discovery of new families of enzymes (Committee on Industrialization of Biology 2015, p. 60). On the dark side, '[c]oncerns for biodiversity are not restricted to wholesale threats to species'; nevertheless, PCSBI (2010) are 'aware of no active or planned research programs involving synthetic biology applied to human genomes' (PCSBI 2010, p. 137). PCSBI's concept of prudent vigilance shares a great deal with the sustainability perspective, as it demonstrates concern for the environment in which future generations will flourish or suffer (Wiek et al. 2012).

2.2.4 Hype Cycle

The discussions above bring to the fore the question of what has bound the spectrum of divergent futures. While both the Royal Academy of Engineering (RAE)'s vision (RAE 2009, p. 7) and EGE's forecast (EGE 2009, p. 24, 31–33) assume and present a linear development of synthetic biology for a singular future, COGEM moderates expectations by referring to the concept of the hype cycle. This cycle generally involves a peak of very optimistic and exaggerated expectations, followed by disappointment and slow recovery phases. Arguing that developments in synthetic biology are currently in the initial stage of the technology hype cycle, COGEM suggest that 'government should first and foremost keep abreast of the latest developments' (COGEM 2008, p. 51). While they expect that 'by providing subsidies and research funding for specific programmes government can respond to expectations and help shape the direction of developments in synthetic biology' (COGEM 2008, p. 51), the promotion of these efforts may 'be modest to avoid hype and thus subsequent disappointment' (Rip 2006, p. 354). Admitting the unpredictability of future developments, they suggest monitoring and trend analysis at an early stage. As a basis of discussions for the report, a COGEM member co-authored a study on the hype cycle in 2010; this study depicts a similar curve between the course of the technology hype cycle and the process of policy formulation and defends the idea that the policy life cycle imposes a delay on the technology hype cycle (Mampuys and Brom 2010; see also Appendix 3 in COGEM 2013). This line of argument then raises the question of how to deal with precautionary policy measures that require formulation prior to the disappointment phase, which can be triggered by wider recognition of risk as in the case of gene therapy (van Lente et al. 2013). Although focusing on different phases of the hype cycle gives different options for projecting a future state of the world (Rip and Robinson 2013), COGEM's report (2008) is silent about how to prepare for the coming disappointment phase of synthetic biology. Strangely enough, it also yields few clues about the link between the precautionary principle and the hype cycle, both of which appear in the same text at various times.

2.2.5 Hype and Exaggerated Claims

The above stagist model not only can limit the scope of technology development and the course of policy action, but also can confine broader societal and ethical discussions. The advocacy of the model is directed at the government, to prevent public discussion from declining with media interest (Mampuys and Brom 2010, p. 167). As a direct consequence, recently, COGEM expressed their concern about hype and exaggerated claims as counterproductive to the development of ethical regulations, while analysing media framing of 'playing God' discourses (COGEM 2013, p. 46–48). The creation of the first organism with a 'synthetic' genome, created by the J. Craig Venter Institute in 2010, provoked discussions on whether we are 'playing God' if we are 'creating life'. The Biotechnology and Biological Sciences Research Council (BBSRC) similarly observe that hype and over-promising in the case of synthetic biology is 'seen as potentially counterproductive to discuss applications and their associated risks and benefits at such an early stage' (BBSRC 2010, p. 76). A German TA newsletter proposes: 'If synthetic biology is to cause specific and new fears, it would need more plausible scenarios than hitherto which would presumably have to refer at least to higher plants or animals' (Sauter 2011, p. 22). A more plausible 'hypothetical, worst-case scenario' can be found in PCSBI's report; that is, 'a newly engineered type of high-yielding blue-green algae cultivated for biofuel production unintentionally leaking from outdoor ponds and out-competing native algal growth' (PCSBI 2010, p. 63); at the same time, they also find the 'playing God' discourse 'to be unhelpful at best, misleading at worst' (PCSBI 2010, p. 156). In contrast, it is clear that before the news of the synthetic genome emerged in 2010, these hype and fear discourses were instead largely welcomed. For instance, an EC report expresses that such discussions should be welcomed but will be productive 'only if we can develop a more sophisticated appreciation of what is meant by "life" than is current in popular discourse' (EC 2005a, p. 19). The European

TA Group (ETAG) states that '[a]nother benefit of these extreme future visions is that they expose the most sensitive issues in the debate and clarify the normative deep core issues at stake' (ETAG 2006, p. iv). As a worst possible scenario, van Est et al. refer to the presence of 'Green goo', comparable with the 'Grey goo' from nanotechnology (van Est et al. 2007, p. 9). More cautiously, EGE (2009) recommend responsible documentation of potential risks and 'hype' benefits in the public debate, 'where the public is confronted, with the assistance of media and science fiction writers, with unrealistic scenarios on synthetic biology products' (EGE 2009, p. 86).

2.3 Discussion

The socio-technical futures depicted in the above TA studies are not independent or separate, but, obviously, interreferential and contextually bounded. This interreferentiality and contextuality suggests how and why the past and present governance of synthetic biology has been locked in the discourse context of the socio-technical future. The following discussion highlights sociocultural and institutional contexts of discourse on the future and governance in the present by contrasting European and American perspectives and institutional TA and other initiatives.

2.3.1 Transatlantic Perspectives

Despite the danger of simplification of complex and dynamic discourses on synthetic biology, a caricature of the contrast between Europe and the United States in this regard may be useful to provoke discussion. In Europe, there is a strong concern for safety in synthetic biology, echoing the GM debate. In the words of Ulrike Felt (2015, p. 3), '"Act now, before it's too late" has become a key-slogan when imagining and performing European futures'. As containment is a basic biosafety strategy, many issues are centred around research practices at laboratory. Where concerned individuals—experts or citizens—hold their own ethical views and attitudes that straggle with the dual use dilemma, a case-by-case and step-by-step procedure is useful to accommodate different values. Normatively holding the precautionary principle, European policy discourses tend in practice to adopt a more manageable approach to synthetic biology under the name of the proportionality principle or of risk-benefit analysis. Such gradual management is in line with the idea of the hype cycle as a stagist model, where efforts to moderate hype and deflate exaggerated claims should make the cycle curve more linear. It is ironic, then, that European discourses on biosecurity threats reflect a conventional understanding of linear, exponential technology and threat trajectory (borrowing from earlier discussions in the US), and that such discourses may themselves appear as exaggerated claims. Their orientation to the future is mostly prognostic, with a 'present future' perspective starting from the present concern for safety.

In the US, concerns about synthetic biology are focused on security, reflecting the perceived threat of terrorism. Discussion of major issues also plays out on a national and global scale, going far beyond individual laboratories. Dual use is a social and institutional matter, and a systematic, spatiotemporally wide view is required. Prudent vigilance, as a synthetic solution to the artificial dichotomy between precautionary and proactionary, seems to convey the idea of preparedness; at the same time, attention to post-incident activities such as response and recovery presumes that possible dangers are not always predictable or preventable in the face of uncertainty. Where human, ecological, and social security risks can be irreversible, the security and sustainability of global natural and innovation ecosystems are underlined in relation to bioeconomy and biodiversity. US discourses exhibit a hermeneutic orientation to the tangible future terror, which then suggests present policy options as seen from the 'future present'.

2.3.2 Institutional TA Initiatives

TA institutions reflexively analyse public concern for biosecurity, biodiversity, and bioeconomy, but do not elaborate these issues or facilitate public debate. The reason may not simply come from the fact that these issues do not match client needs for a single TA institution to address, but rather reflect the relevant experience of the GM safety debate. As observed in 2.2.3, non-TA science/ethics advisory institutions, like BBSRC, COGEM, EGE, and CBE and CNEV, make no secret of their concerns about biodiversity issues in synthetic biology. Under these circumstances, it is curious how European TA institutions discuss gene drive technology, an area in which genome editing has been rapidly growing and where many social and policy discussions remain focused on the ethical implications of its medical applications to human embryos (cf. Giese 2017).

After the news of the synthetic genome broke in 2010, many policy documents pointed out that hype and exaggerated claims are rather counterproductive for the development of technologies, applications, and regulations, proposing more plausible scenarios (BBSRC 2010; PCSBI 2010; Sauter 2011; COGEM 2013). In contrast, before the release of the synthetic genome in 2010, TA institutions (ETAG 2006; van Est et al. 2007) were rather positive about proposing extreme future visions and worst cases. It is interesting to contrast them with Europe-wide institutions, which were more reflexive and cautious about presenting extreme and exaggerated claims even before 2010, expressing some reservations about whether worst-case scenarios had a 'sophisticated appreciation of what is meant by "life"' (EC 2005a), or constituted 'responsible reporting' (EGE 2009). TA institutions thus clearly shifted their orientations to techno-futures involving utopian or dystopian hermeneutics (that is, from mode [b]) to normative and purpose-related hermeneutics (mode [a]), though other institutions changed their projections of bounded divergence of futures less from before to after this disruptive innovation in synthetic biology.

3   Conclusion

This study has demonstrated how TA discourses are politicized and socio-technical discursive futures are affected by sociocultural and institutional contexts. The reflexive hermeneutic approach applied in the case study of synthetic biology has revealed that future-oriented conceptions of the 'precautionary' and 'proactionary' are not mere political contrasts, but rather respectively intertwined with general concepts including biosafety, biosecurity, bioeconomy, and biodiversity. Hype and exaggerated claims in perceptions of possible futures have also been affected by these general concepts and the advent of disruptive innovation. This 'present future' perspective on the future has not much shifted present practices and processes in technology governance but rather follows them. TA institutions are thus required to reflexively analyse how their own discourses are based on 'present future' or 'future present' perspectives and explore how they can open future spaces going against the hermeneutic imperialism of science.

Acknowledgements This work was fully supported by the European Academy of Technology and Innovation Assessment (EA), in Germany, during my time there as an invited research fellow in Autumn 2015. Comments from Armin Grunwald, Loet Leydesdorff, Petra Ahrweiler and Stephan Lingner at the Uncertain Futures Workshop at EA and thereafter were particularly valuable for the development of earlier versions of the manuscript.

References

Adam, B., & Groves, C. (2007). *Future matters: Action, knowledge, ethics*. Leiden: Brill.
Angermüller, J. (2011). Heterogeneous knowledge: Trends in German discourse analysis against an international background. *Journal of Multicultural Discourses, 6*(2), 121–136.

Cantley, M., & Lex, M. (2011). Genetically modified foods and crops. In J. B. Wiener, M. D. Rogers, J. K. Hammitt, & P. H. Sand (Eds.), *The reality of precaution: Comparing risk regulation in the US and Europe* (pp. 39–64). Washington: RFF Press.

Collingridge, D. (1980). *The social control of technology*. London: Pinter.

Committee on Research Standards and Practices to Prevent the Destructive Application of Biotechnology. (2004). *Biotechnology research in an age of terrorism*. Washington, D.C.: National Academies Press.

Cronberg, T. (1996). European TA-discourses: European TA? *Technological Forecasting and Social Change, 51*(1), 55–64.

Cunliffe, A. L. (2011). Crafting qualitative research: Morgan and Smircich 30 years on. *Organizational Research Methods, 14*(4), 647–673.

deLeon, P. (1988). *Advice and consent: The development of the policy sciences*. New York: Sage.

Dryzek, J. S. (1982). Policy analysis as a hermeneutic activity. *Policy Sciences, 14*(4), 309–329.

Durning, D. (1999). The transition from traditional to postpositivist policy analysis: A role for Q-methodology. *Journal of Policy Analysis and Management, 18*(3), 389–410.

EGE-European Group on Ethics. (2000). *Opinion 15. Ethical aspects of stem cell research and use*. Brussels: European Commission.

EGE-European Group on Ethics. (2003). *Opinion 17. Opinion on the ethical aspects of clinical research in developing countries*. Brussels: European Commission.

Ely, A., Van Zwanenberg, P., & Stirling, A. (2014). Broadening out and opening up technology assessment: Approaches to enhance international development, co-ordination and democratisation. *Research Policy, 43*(3), 505–518.

European Commission. (2012). *Innovating for sustainable growth: A bioeconomy for Europe*. Luxembourg: Publications Office of the European Union.

Felt, U. (2015). *The temporal choreographies of participation: Thinking innovation and society from a time-sensitive perspective. Pre-print*. Vienna: University of Vienna, Department of Science and Technology Studies.

Fischer, F. (1998). Beyond empiricism: Policy inquiry in postpositivist perspective. *Policy Studies Journal, 26*(1), 129–146.

Fischer, F. (2003). *Reframing public policy: Discursive politics and deliberative practices*. Oxford: Oxford University Press.

Forge, J. (2010). A note on the definition of 'dual use'. *Science and Engineering Ethics, 16*(1), 111–118.

Freeman, C., & Perez, C. (1988). Structural crises of adjustment, business cycles and investment behaviour. In G. Dosi, C. Freeman, R. Nelson, G. Silverberg, & L. Soete (Eds.), *Technical change and economic theory* (pp. 38–66). London: Pinter.

Gibbons, M. T. (2008). Hermeneutics. In W. A. J. Darity Jr. (Ed.), *International encyclopedia of the social sciences* (2nd ed.). Detroit: Thompson Sale.

Giese, B. (2017). *Gene drives-a new quality in GMO release*. Paper presented at 3rd European Technology Assessment Conference, Cork, 16 May 2017.

Grunwald, A. (2014a). Modes of orientation provided by futures studies: Making sense of diversity and divergence. *European Journal of Futures Research, 2*, 30.

Grunwald, A. (2014b). The hermeneutic side of responsible research and innovation. *Journal of Responsible Innovation, 1*(3), 274–291.

Hermerén, G. (2012). The principle of proportionality revisited: Interpretations and applications. *Medical Health Care and Philosophy, 15*(4), 373–382.

Hitzler, R. (2005). The reconstruction of meaning: notes on German interpretive sociology. *Forum: Qualitative Social Research, 6*(3), Art. 45.

Jasanoff, S. (2003). Technologies of humility: Citizen participation in governing science. *Minerva, 41*(3), 223–244.

Kaebnick, G. E., Gusmano, M. K., & Murray, T. H. (2014). The ethics of synthetic biology: Next steps and prior questions. *Hastings Center Report, 44*(6), S4–S26.

Keasling, J. (2010). Testimony to Committee on Energy and Commerce. *U.S. House of representatives*, 27 May 2010.

Keller, R. (2005). Analysing discourse: An approach from the sociology of knowledge. *Forum: Qualitative Social Research, 6*(3), Art. 32.

Keller, R. (2011). The sociology of knowledge approach to discourse. *Human Studies, 34*(1), 43–65.

Keller, R. (2012). Entering discourses: A new agenda for qualitative research and sociology of knowledge. *Qualitative Sociology Review, 8*(2), 46–75.

Kuhlau, F., Eriksson, S., Evers, K., & Höglund, A. T. (2008). Taking due care: Moral obligations in dual use research. *Bioethics, 22*(9), 477–487.

Lentzos, F., & Rose, N. (2009). Governing insecurity: Contingency planning, protection, resilience. *Economy and Society, 38*(2), 230–254.

Li, F., Owen, R., & Simakova, E. (2015). Framing responsible innovation in synthetic biology: The need for a critical discourse analysis approach. *Journal of Responsible Innovation, 2*(1), 104–108.

Lingner, S. (2017). Imagining socio-technological futures—Lessons from past visions. In D. M. Bowman, et al. (Eds.), *The politics and situatedness of emerging technologies* (pp. 39–50). Berlin: AKA/IOS Press.

Luhmann, N. (1982). *The differentiation of society*. New York: Columbia University Press.

Majone, G. (1977). Technology assessment and policy analysis. *Policy Sciences, 8*(2), 173–175.

Mampuys, R., & Brom, F. W. A. (2010). The quiet before the storm: Anticipating developments in synthetic biology. *Poiesis & Praxis: International Journal of Ethics of Science and Technology Assessment, 7*(3), 151–168.

Marris, C., Jefferson, C., & Lentzos, F. (2014). Negotiating the dynamics of uncomfortable knowledge: The case of dual use and synthetic biology. *BioSocieties, 9*(4), 393–420.

McLeish, C., & Nightingale, P. (2007). Biosecurity, bioterrorism and the governance of science: The increasing convergence of science and security policy. *Research Policy, 36*(10), 1635–1654.

Morçöl, G. (2002). *A new mind for policy analysis: Toward a post-Newtonian and post-positivist epistemology and methodology*. Westport: Praeger.

OECD. (1986). Recombinant DNA safety considerations. Safety considerations for industrial, agricultural and environmental applications of organisms derived by recombinant DNA techniques. http://www.oecd.org/sti/emerging-tech/40986855.pdf.

Parens, E., Johnston, J., & Moses, J. (2009). *Ethical issues in synthetic biology: An overview of the debates. SYNBIO 3, June 2009*. Washington, D.C.: Woodrow Wilson International Center for Scholars.

Pauwels, E., & Ifrim, I. (2008). *Trends in American & European press coverage of synthetic biology: Tracking the last five years of coverage. SYNBIO 1, November 2008.* Washington, D.C.: Woodrow Wilson International Center for Scholars.

Rabinow, P., Bennett, G., & Stavrianakis, A. (2006). Response to "Synthetic genomics: Options for governance". *ARC Concept Note*, No. 10, 5 December 2006.

Rath, J. (2014). Rules of engagement: Restricting security-sensitive research data-A European view. *EMBO Reports, 15*(11), 1119–1122.

Reichertz, J. (2004). Objective hermeneutics and hermeneutic sociology of knowledge. In U. Flick, E. von Kardorff, & I. Steinke (Eds.), *A companion to qualitative research* (pp. 290–295). Thousand Oaks: Sage.

Rein, M., & White, S. H. (1977). Can policy research help policy? *Public Interest, 49,* 119–136.

Rip, A. (1986). Controversies as informal technology assessment. *Knowledge: Creation, Diffusion, Utilization, 8*(2), 349–371.

Rip, A. (2006). Folk theories of nanotechnologists. *Science as Culture, 15*(4), 349–365.

Rip, A., & Robinson, D. K. R. (2013). Constructive technology assessment and the methodology of insertion. In N. Doorn, D. Schuurbiers, I. van de Poel, & M. E. Gorman (Eds.), *Early engagement and new technologies: Opening up the laboratory* (pp. 37–53). Berlin: Springer.

Sheppard, A. W., Gillespie, I., Hirsch, M., & Begley, C. (2011). Biosecurity and sustainability within the growing global bioeconomy. *Current Opinion in Environmental Sustainability, 3*(1–2), 4–10.

Stirling, A. (2008). "Opening up" and "closing down": Power, participation, and pluralism in the social appraisal of technology. *Science, Technology, and Human Values, 33*(2), 262–294.

Torgerson, D. (1986). Between knowledge and politics: Three faces of policy analysis. *Policy Sciences, 19*(1), 33–59.

van den Ende, J., Mulder, K., Knot, M., Moors, E., & Vergragt, P. (1998). Traditional and modern technology assessment: Toward a toolkit. *Technological Forecasting and Social Change, 58*(1–2), 5–21.

van Eijndhoven, J. (1997). Technology assessment: Product or process? *Technological Forecasting and Social Change, 54*(2–3), 269–286.

van Lente, H., Spitters, C., & Peine, A. (2013). Comparing technological hype cycles: Towards a theory. *Technological Forecasting and Social Change, 80*(8), 1615–1628.

Vig, N. J., & Paschen, H. (Eds.). (2000). *Parliaments and technology: The development of technology assessment in Europe.* New York: State University of New York Press.

Vogel, K. M. (2008). Framing biosecurity: An alternative to the biotech revolution model? *Science and Public Policy, 35*(1), 45–54.

von Schomberg, R. (2006). The precautionary principle and its normative challenges. In E. Fisher, J. Jones, & R. von Schomberg (Eds.), *Implementing the precautionary principle: Perspectives and prospects* (pp. 19–41). Cheltenham: Edward Elgar.

White House (2012). *National bioeconomy blueprint.* April 2012.

Wodak, R. (2011). Complex texts: Analyzing, understanding, explaining and interpreting meanings. *Discourse Studies, 13*(5), 623–633.

Wiek, A., Guston, D., Frow, E., & Calvert, J. (2012). Sustainability and anticipatory governance in synthetic biology. *International Journal of Social Ecology and Sustainability Development, 3*(2), 25–38.
Wiener, J. B., Rogers, M. D., Hammitt, J. K., & Sand, P. H. (Eds.). (2011). *The reality of precaution: comparing risk regulation in the US and Europe.* Washington, D.C.: RFF Press.
Wynne, B. (1975). The rhetoric of consensus politics: A critical review of technology assessment. *Research Policy, 4*(2), 108–158.
Wynne, B. (1993). Public uptake of science: A case for institutional reflexivity. *Public Understanding of Science, 2*(4), 321–337.
Wynne, B. (2014). Further disorientation in the hall of mirrors. *Public Understanding of Science, 23*(1), 60–70.
Wynne, B. (2016). Ghosts of the machine: Publics, meanings and social science in a time of expert dogma and denial. In J. Chilvers & M. Kearnes (Eds.), *Remaking participation: Science, environment and emergent publics* (pp. 99–120). London: Routledge.

List of Analysed Documents

Biotechnology and Biological Sciences Research Council (BBSRC). (2010). Engineering and Physical Sciences Research Council (EPSRC) and Sciencewise-ERC. Synthetic Biology Dialogue. https://bbsrc.ukri.org/documents/1006-synthetic-biology-dialogue-pdf/.
Bunn, S. (2008). *Synthetic biology, postnote 298.* London: Parliamentary Office of Science and Technology (POST).
Committee on Industrialization of Biology. (2015). A roadmap to accelerate the advanced manufacturing of chemicals, Board on Chemical Sciences and Technology, Board on Life Sciences; Division on Earth and Life Studies & National Research Council. Industrialization of Biology: A Roadmap to Accelerate the Advanced Manufacturing of Chemicals. Washington, D.C.: National Academies Press.
de Vriend, H. (2006). *Constructing life: Early social reflections on the emerging field of synthetic biology. Working document 97.* The Hague: Rathenau Instituut.
European Academies Science Advisory Council (EASAC). (2010). Realising European potential in synthetic biology: Scientific opportunities and good governance. EASAC policy report, 13. Halle: German Academy of Sciences Leopoldina. https://www.cbd.int/doc/emerging-issues/emergingissues-2013-10-EASAC-SyntheticBiology-en.pdf.
European Commission (EC). (2005a). *Synthetic biology-Applying engineering to biology: Report of a NEST high-level expert group.* Project report EUR 21796. Luxembourg: Office for Official Publications of the European Communities.
European Commission (EC). (2005b). *Reference document on synthetic biology. 2005/2006-NEST-PATHFINDER INITIATIVES.* Brussels: European Commission.
European Group on Ethics in Science and New Technologies (EGE). (2009). Ethics of synthetic biology. In Opinion of the European Group on ethics in science and new technologies to the European Commission, No. 25. http://ec.europa.eu/archives/bepa/european-group-ethics/docs/opinion25_en.pdf.

European Parliamentary Technology Assessment (EPTA). (2011). Synthetic biology. Briefing note 1.

European Technology Assessment Group (ETAG). (2006). *Technology assessment on converging technologies. IP/A/STOA/ST/2006-6.* Brussels: European Parliament.

Garfinkel, M. S., Endy, D., Epstein, G. L., & Frieman, R. M. (2007). Synthetic genomics: Options for governance. J. Craig Venter Institute, Center for Strategic and International Studies & Massachusetts Institute of Technology's Department of Biological Engineering. https://www.bio.org/sites/default/files/synthetic-genomics-report.pdf.

Health Council of the Netherlands (GR), Advisory Council on Health Research (RGO), & Royal Netherlands Academy of Arts and Sciences (KNAW). (2008). *Synthetic biology: Creating opportunities, 19E.* The Hague: Health Council of the Netherlands.

International Risk Governance Council (IRGC). (2010). *Policy brief: Guidelines for the appropriate risk governance of synthetic biology.* Lausanne: IRGC.

Joyce, S., Mazza, A-M., & Kendall, S. (2013). Positioning synthetic biology to meet the challenges of the 21st century-Summary report of a six academies symposium series. Committee on Science, Technology, and Law Policy and Global Affairs, Board on Life Sciences Division on Earth and Life Studies & National Academy of Engineering. Washington, D.C.: National Academies Press.

National Science Advisory Board for Biosecurity (NSABB). (2010). Addressing biosecurity concerns related to synthetic biology. https://osp.od.nih.gov/wp-content/uploads/NSABB_SynBio_DRAFT_Report-FINAL-2_6-7-10.pdf.

Nuffield Council on Bioethics (NCB). (2012). *Emerging biotechnologies: Technology, choice and the public good.* London: Nuffield Council on Bioethics.

Organisation for Economic Co-operation and Development (OECD). (2014). *Emerging policy issues in synthetic biology.* Paris: OECD Publishing.

Organisation for Economic Co-operation and Development (OECD), & Royal Society (RS). (2010). Symposium on opportunities and challenges in the emerging field of synthetic biology: Synthesis report. https://www.oecd.org/science/emerging-tech/45144066.pdf.

Presidential Commission for the Study of Bioethical Issues (PCSBI). (2010). New directions: The ethics of synthetic biology and emerging technologies. https://bioethicsarchive.georgetown.edu/pcsbi/synthetic-biology-report.html.

Rerimassie, V., & Stemerding, D. (2014). *SynBio politics-Bringing synthetic biology into debate.* The Hague: Rathenau Instituut.

Royal Academy of Engineering (RAE). (2009). *Synthetic biology: Scope, applications and implications.* London: Royal Academy of Engineering.

Sauter, A. (2011). *Synthetic biology: Final technologisation of life-Or no news at all? TAB-Brief 39.* Berlin: Büro für Technikfolgen-Abschätzung beim Deutschen Bundestag (TAB).

Spanish Bioethics Committee (CBE), & Portuguese National Ethics Council for the Life Sciences (CNECV). (2011). Synthetic biology. Joint report. http://www.cnecv.pt/admin/files/data/docs/1335183813-cnecv-cbe-synthetic-biolo.pdf.

Stemerding, D., & Rerimassie, V. (2013). *Discourses on synthetic biology in Europe.* The Hague: Rathenau Instituut.

SYBHEL. (2012). Synthetic biology & human health: Ethical and legal issues. Final report & policy recommendations. https://cordis.europa.eu/result/rcn/147536_en.html.

SYNBIOSAFE. (2008). Publishable executive summary.

SYNTH-ETHICS. (2011). Ethical and regulatory challenges raised by synthetic biology. Final report summary. https://cordis.europa.eu/project/rcn/90989_en.html.

The National Academies Keck Futures Initiative (NAFKI). (2010). *Synthetic biology: Building a nation's inspiration-Interdisciplinary research team summaries.* Washington, D.C.: National Academies Press.

The Netherlands Commission on Genetic Modification (COGEM). (2008). Biological machines? Anticipating developments in synthetic biology. COGEM report, CGM/080925-01.

The Netherlands Commission on Genetic Modification (COGEM). (2013). Synthetic biology-Update 2013: Anticipating developments in synthetic biology. COGEM topic report, CGM/130117-01.

UK Synthetic Biology Roadmap Coordination Group. (2012). *A synthetic biology roadmap for the UK.* Swindon: Technology Strategy Board.

van Est, R., & Stemerding, D. (Eds.). (2012). Making perfect life: European governance challenges in 21st century bio-engineering. Final report, IP/A/STOA/FWC/2008-096/LOT6/C1/SC3. Brussels: European Parliament, Science and Technology Options Assessment (STOA).

van Est, R., de Vriend, H., & Walhout, B. (2007). *Constructing life: The world of synthetic biology.* The Hague: Rathenau Instituut.

Vedelsby, J. (2011a). Synthetic biology: Challenges and debates-New developments within biotechnology and genetic engineering, Fra rådet til tinget, 281.

Vedelsby, J. (2011b). Synthetic biology: A discussion paper. Danish Council of Ethics and Danish Board of Technology. http://www.etiskraad.dk/~/media/Etisk-Raad/en/Publications/Synthetic-biology-2011.pdf?la=da.

Go Yoshizawa (Ph.D.) is research fellow for the project "Positive Environment in Public Participation and Engagement for Responsible Research and Innovation" (PEPPER) at Work Research Institute (AFI), Oslo Metropolitan University.

In vitro Meat: The Normative Power of a Vision in the Innovation and Transformation Process

Inge Böhm, Silvia Woll and Arianna Ferrari

Abstract

The current production of meat and the high meat consumption have negative effects on humans, animals and the environment. A sustainable reorientation of mass production and mass consumption is not possible. However, a technological solution is in sight: in vitro meat is presented as an innovation which may solve the problems of today's meat production and consumption without giving up on meat. We interviewed experts and stakeholders and confronted them with the vision of this innovation. The results of these interviews provide an insight in current visions and imaginations of possible futures that shows which particular challenges result from assessing the emerging technology in vitro meat, also beyond in vitro meat itself.

I. Böhm (✉)
Mannheim, Germany
e-mail: boehm.inge@gmx.net

S. Woll
Institute for Technology Assessment and Systems Analysis, Karlsruhe Institute of Technology, Karlsruhe, Germany
e-mail: silvia.woll@kit.edu

A. Ferrari
Department Strategy and Content, Futurium gGmbH, Berlin, Germany
e-mail: ferrari@futurium.de

© Springer Fachmedien Wiesbaden GmbH, part of Springer Nature 2019   211
A. Lösch et al. (eds.), *Socio-Technical Futures Shaping the Present*,
Technikzukünfte, Wissenschaft und Gesellschaft / Futures of Technology,
Science and Society, https://doi.org/10.1007/978-3-658-27155-8_10

1   Introduction

One of the most urgent problems of our time concerns the question of how mankind will be able to feed itself. The search for solutions for feeding a growing
world population reveals that the way in which we produce and consume meat
today is extremely inefficient and in many respects unsustainable. The current
production of meat and the high meat consumption affect human health, the
well-being of animals as well as the environment and a sustainable reorientation
of mass production and mass consumption is not possible (e.g. Albritton 2013;
Heinrich-Böll-Stiftung & BUND 2016; Wellesley et al. 2015). However, a technological solution is in sight: in vitro meat is presented as an innovation which
may solve the problems of today's meat production and meat consumption without giving up on meat.

   Since the first beef burger made of in vitro meat was fried at a public tasting
in August 2013 and provided the proof of concept for this technological innovation, according to statements made by the innovators from in vitro meat research
there exists the possibility to sustainably produce large quantities of meat and
thus to reduce or even eliminate the negative effects on humans, animals and the
environment. Innovators argue that in vitro meat does not only provide a plausible
technological solution for current problems of today's meat production and meat
consumption, but that it is also the most realistic solution because humans will
not stop eating meat (Bhat et al. 2015; Post 2014b).

   The project "Visionen von In vitro-Fleisch. Analyse der technischen und ges-
amtgesellschaftlichen Aspekte und Visionen von In vitro-Fleisch" (Visions of
in vitro meat. Analysis of the technical and social aspects and visions of in vitro
meat-VIF) aimed at analysing the role of visions for innovation and transfer processes as well as their governance. It was funded by the Federal Ministry of Education and Research (BMBF) in the context of its "Initiative Innovations- und
Technikanalyse" (Innovation and Technology Analysis initiative-ITA). The project analysed the social, cultural and political aspects of the visions of the current in vitro meat research and the final goal was to formulate research-political
options for the Ministry.

   Being still a proof of concept, in vitro meat interacts with society in form of
a vision, i.e. as a representation of a future world. Vision assessment has been
developed in TA to answer the need of evaluating representations of the future in
innovation processes (Grin and Grunwald 2000). Vision assessment has been followed by the hermeneutic approach to visions, which has focused on the content

and significance of visions.[1] In the VIF project we followed a more empirical approach, which was elaborated in the context of a related basic research project at the Institute for Technology Assessment and Systems Analysis (ITAS) at the Karlsruhe Institute of Technology (KIT).[2] Visions are investigated as socio-epistemic practices capable of acting on discourses, providing orientation for activities and producing new social orders and thus influencing innovation.

In this contribution we would like to disentangle the dynamics of promises, expectations and the activities generated by them, exploring the innovators' vision on in vitro meat and the responses of different experts and stakeholders, examined through interviews. In the course of the VIF project, a total number of twelve semi-structured interviews were conducted in 2016 to analyse the general principles and visions of today's in vitro meat research and to investigate scientific, technological, social, cultural and political aspects, mainly in Germany. Five experts from the fields of tissue engineering, bio-engineering, in vitro meat research and food technology were interviewed as well as seven stakeholders from the fields of animal rights, environment protection, politics, the food industry, food service and ecological and conventional agriculture. We wanted to create both societal and political courses of action in dialogue with different actors and the public.

After having shown how different experts and stakeholders reacted to the innovators' vision of in vitro meat by developing their own visions on different aspects of this innovation (namely its technical feasibility and its ethicality), we will demonstrate how the engagement with this innovators' vision generated different responses in Germany about the desirability of in vitro meat. Finally we will focus on what we can learn from interviewing experts and stakeholders about futures which shape the present assessment of an innovation.

2   The Innovators' Vision of In vitro Meat

In vitro meat is presented by the innovators as *the* solution to all existing problems of today's meat production and consumption, being, they say, sustainable, ethical and having the potential to meet mankind's growing demand for meat. Bhat et al. (2015) even present it as the inevitable future of mankind.

---

[1]For a more exhaustive discussion of these methods see Grunwald (2004), (2012), (2015).
[2]See https://www.itas.kit.edu/projekte_loes14_luv.php. The approach developed in the project goes beyond the method of classical vision assessment and includes the analysis of the practical effects of such visions on development processes and innovation dynamics.

"We are developing cultured animal products with no animal slaughter and much lower use of land, water, energy, and chemicals." (Modern Meadow 2016)

"The world faces critical food shortages in the near future as demand for meat is expected to increase by more than two-thirds, according to the Food and Agriculture Organization of the United Nations. (…) Cultured Beef represents the crucial first step in finding a sustainable alternative to meat production." (Cultured Beef 2017a)

"Our vision is a strong foundation of accessible, public, fundamental cellular agriculture research, upon which we can build a post-animal bioeconomy, where we harvest animal products from cell cultures, not animals, to feed a growing global population sustainably and affordably." (New Harvest 2017)

"In vitro meat production is the inescapable future of humanity. (…) In the light of sizable negative effects of current meat production on environment and human health, a viable solution lies with in vitro meat production, a process that poses to revolutionize human existence." (Bhat et al. 2015)

According to this way of arguing, it is scientifically proven that it is impossible to organise conventional meat production and the consumption of meat in a sustainable way, as both produce many negative effects on the environment, animals, climate change, health etc. It is thus necessary to find alternatives to conventional meat production. There are indeed other protein sources, such as insects, seaweed or plants. A vegan or vegetarian diet is considered the best option for solving the problems. From the point of view of the innovators, however, this is not feasible. People will not become vegans, vegetarians or insect eaters. The innovators believe that most people love meat and will not stop eating it, for whatever reason. In vitro meat is considered the alternative to conventionally produced meat which is most likely to replace today's meat. Thus, in vitro meat is considered the most realistic, most sustainable alternative to conventional meat production. Some authors have even drawn the conclusion that we might be morally obliged to eat in vitro meat (Bhat and Bhat 2010; Schaefer and Savulescu 2014).

In the case of in vitro meat, technology assessment (TA) is in particular confronted with ideas of the future, since in vitro meat, being an innovation still in its infancy, is much based on expectations, promises and the construction of future. From the point of view of TA as the research of *consequences*, however, assessments of the feasibility or desirability of such *technological futures* (technology-related ideas of the future) look little meaningful or even risky. In the course of dealing with new and emerging science and technologies (NEST) this has increasingly become obvious, and the method of vision assessment (Grin and Grunwald 2000) has provided a first answer. As yet unanswered is the question of

TA knowledge generation, methods and theories when it comes to dealing with current technological futures. The ITAS/KIT project "Leitbilder und Visionen als sozio-epistemische Praktiken-Theoretische Fundierung und praktische Anwendung des Vision Assessments in der Technikfolgenabschätzung" (Visions as socio-epistemic practices-Theoretical foundation and practical application of Vision Assessment in Technology Assessment) which provides a theoretical basis for the analysis of the in vitro meat innovation deals with finding an answer to this question. The reason for this project was, among others, the critical debate on Grunwald's *hermeneutic TA* as a decoding of attributions of meaning of future technologies by way of current technological futures.

The analysis of visionary ideas is not sufficient for the analysis and assessment of the process effectiveness of visions. TA needs knowledge of the effectiveness of these visionary ideas to be able to, as an accompaniment of currently running processes, make statements on the functions of visions within dynamic socio-technological constellations. Thus, working out the theoretical concept of visions resulted in defining visions as socio-epistemic practices which generate meanings, provide orientation for activities and are capable of producing new social orders. From this point of view, visions are ideas of the future whose practical effectiveness can be analysed according to specific functions.

In their article "How Smart Grid Meets In Vitro Meat: on Visions as Socio-epistemic Practices" (Ferrari and Lösch 2017), Ferrari and Lösch present four functions of visions which have emerged from the analysis of two technologies, namely smart grids and in vitro meat. In relation to in vitro meat it has been pointed out:

1. *"Visions serve as interfaces allowing for transitions between presence and future"* (Ferrari and Lösch 2017). In vitro meat allows for a transition from current problems of meat production and consumption to a sustainable diet by help of in vitro meat. The innovators present in vitro meat as an environment-friendly alternative, thus referring to hypothetical, selected scenarios which are presented as predictions. They do not communicate, however, that these scenarios are based on insecure assumptions because as yet there is no large-scale production process and the anticipative life cycle analyses show several weak spots and are less conclusive. Alternative solutions for currently existing problems are marginalised or dismissed as being unrealistic. Risks are belittled as challenges which can be solved.

2. *"Visions work as communication media between different discourses and integrate different interests"* (Ferrari and Lösch 2017). This was illustrated by the burger made of bovine stem cells presented by Mark Post and his team

at Maastricht University in 2013. Here, a scientific object, or more exactly: a product of tissue engineering, is transformed into a possibility to realise a sustainable diet. Due to the communicative power of the vision, there happens a cross-over of the topics of the different actors who up to then had independently dealt with the topic of sustainability. Among these topics are the concern about the earth's ecological conditions or the concern about our ways of treating animals while, at the same time, praising meat as being fundamental for human nutrition. In this way the innovators are able to make in vitro meat a part of the sustainability discourse.

3. *"Visions allow for coordinating several activities"* (Ferrari and Lösch 2017). In vitro meat creates new alliances of previously competing actors. These activities have presented in vitro meat as a possible solution for food policy: in vitro meat enters the stages of several sustainability debates and is gradually perceived as one option among others. These new alliances become obvious by the crossing-over of the arguments of animal rights activists and in vitro meat researchers as well as of conventional and ecological agricultural associations, of environment protectors and politics.

4. *"Visions unfold a normative power"* (Ferrari and Lösch 2017). The idea that in vitro meat is the best and only solution for the environment, humans and animals is construed by marginalising the alternative solution of plant-based food and leaving the future of animals open (Ferrari 2016). Although the innovators present vegetarianism or a reduction of meat consumption as the best solution for the environment, by referring to the crucial significance of meat in the history of mankind they declare this option not feasible. The situation of future meat consumption, whose uncertainty results from the fact that the current methods and dimensions of the production and consumption of meat will no longer be possible and must be changed, seems to be clarified by the announcement that in vitro meat is the future solution for all meat eaters.

Currently there is still a mood of optimism, the innovators present in vitro meat as the only realistic solution to the problems of current meat production and meat consumption, although some scepticism has developed, especially related to the presumed ethical and social advantages of this innovation. During the first half of the project we noticed that there were hardly any sceptical voices in the scientific and media discourse which might critically discuss the technology of in vitro meat. We therefore chose to investigate experts' and stakeholders' assessments of this innovation by confronting them with the innovators' vision on in vitro meat, giving them also space to develop their own visions of in vitro meat. Interestingly enough, particularly in the cases of ethical and social aspects of this innovation, the engagement with the innovators' vision also generated official statements and responses by different actors in Germany. We are not arguing here that these statements are direct consequences of the activity of our project. However, the visibility of in vitro meat gained in Germany through the activities of the innovators, the media coverage and the scientific activities of our project contributed to enriching the debate and stimulating a more diversified assessment of this innovation.

3   Assessing the Technical Feasibility of In vitro Meat

The idea of in vitro meat is not at all new. It was already formulated in a much referred to quotation by Winston Churchill in the year 1931: "We shall escape the absurdity of growing a whole chicken in order to eat the breast or wing, by growing these parts separately under a suitable medium" (Churchill 1931). Meat which is fit for human consumption consists mostly of animal muscular tissue which may also be produced outside the body (ex vivo), by way of culture cells. In the research context, the thus applied method of tissue engineering is predominantly used by medicine for the reproduction of destroyed tissue or malfunctioning cells and organs. By way of a biopsy, muscle stem cells are extracted and bred in vitro in the laboratory. Depending on the kind of cells, they may be cultured two-dimensionally or three-dimensionally with the help of frames. These cells are cultivated in a nutrition medium where they breed (proliferation). In the course of the process of differentiation, finally the stem cells become muscle cells (myoblasts), the mono-nuclear myoblasts grow into multi-nuclear myotubes and then form myofibrils or muscle fibres (myogenesis). About 20,000 of these fibres are needed to make one hamburger patty (Cultured Beef 2017b).

"The development of cultured beef from bovine skeletal muscle stem cells through tissue-engineering techniques is potentially a resource-efficient way to grow meat. Proof of concept was provided in 2013 by our lab, yet it is fair to say that the technology is still in its infancy and that some obstacles need to be cleared before it can reach its full potential." (Post 2014a)

Currently the large-scale production of in vitro meat is not possible. Above all, it requires additional research on suitable cells or types of cells, nutrition medium and nutrient composition, suitable material for the scaffolds as well as guaranteed cell metabolism. During our interviews for the VIF project, experts in this innovation or in related areas pointed out further interesting aspects: The development of efficient bio-reactors of sufficient size is necessary (see Interviews C, I, J). There is still a lack of different components like cultured fat or the need for making the cells produce blood or myoglobin. Researchers are already working on this, but it will take "a few years" to develop (see Interview C). Also the chemical definition of the media is not yet entirely known. Cell lines from mice are currently used in some research which an in vitro meat researcher considers potentially unsafe for food because cell lines are genetically engineered and it is not entirely known how this interacts with the human body (see Interview I); also in vitro meat from these cell lines would generate mouse meat (see Interview J). A tissue engineer does not expect it to be possible that in vitro meat might be produced on a large scale (see Interview J). So the evaluation among experts of the feasibility to grow meat on a large scale widely varies: "It often sounds very good, the idea of in vitro meat. But at a closer look there are many obstacles and ethical considerations." (Interview J)

Most of the expert interviews aim at shedding light on as yet little discussed aspects of the research on in vitro meat which are only known to specialists. According to the experts' statements, there is still a considerable need for basic research, and there are problems when it comes to transferring the results produced by tissue engineering. For an innovation at an early stage of development, neither these results nor the fact that innovators tend to be very optimistic in their previsions are unusual or surprising. However, what emerged was the fact that the experts who were sceptical about the feasibility of in vitro meat were the ones pointing out all the technical difficulties and the need for much funding in research more easily, while at the same time seriously doubting that research on this innovation was meaningful.

The technical feasibility of in vitro meat does not only have to do with the biological challenges, but also with the opportunity costs of this innovation. The highest attention is given to the ecological advantages: therefore the methods of life cycle assessment or rather of anticipated life cycle assessment play a central role for the TA of this innovation.

According to some previously published life cycle assessment analyses, the production of in vitro meat would require less agricultural acreage, it would consume less water and it would emit less pollutants as well as less greenhouse gas than the breeding of animals for producing an adequate amount of food (Tuomisto 2011; Mattick et al. 2015). However, it might require more energy for the bioreactors, which is often produced by burning fossil energy sources, than needed for the production of pork and poultry.

During our interviews, an in vitro meat researcher estimated that the energy consumption of in vitro meat "will still be quite significant", but in a bioreactor as a closed system "you have a lot of potential to recycle energy, to isolate the bioreactor into recycling the energy you use" (Interview C). Another in vitro meat researcher expects that research will "be able to optimise the system and reduce power consumption from what has been projected", but reminds that the "bioreactor for a certain class of the cultured meat production again is not fully defined. [...] So how you can project the energy it's going to need is very difficult."

These uncertainties pointed out by the experts confirmed the general challenges of anticipated life cycle assessments in TA. Since there is still no established method to produce in vitro meat on a large-scale basis, the possible ecological consequences remain hypothetical and built on scenarios. Furthermore it has become evident that these studies are still too few in order to make robust projections (cf. Mattick et al. 2015). Most importantly these interviews showed that the interpretation of the in vitro meat trends, i.e. its future, really depend much on the general attitude toward this innovation: the experts in favour of in vitro meat tended to see the problems to be solved as challenges while underlying the potential of this innovation.

"With any new technology, the cost will come down, the energy use will come down as we optimise the system. So then, I think we should be under no illusion that cultured meat is going to be that one hundred million times better in every single way or form because I think that that would be kind of misinforming people, too." (Interview I)

The experts who are generally more critical toward the very idea of in vitro meat see the challenges as obstacles which are very difficult to overcome. A tissue engineer pointed out the problem of high energy consumption for bioreactors:

"If one would create an energy balance on that, that would be horrifying." (Interview J).

Despite the fact that the technical difficulties have not yet been solved, there is a lack of scientific critique of in vitro meat. Further LCA studies have not been published (the newest paper by Alexander et al. (2017) reports the results of the previous LCA studies) and the experts with critical voices toward this innovation are not present in the public debate. So far, the innovators' vision on in vitro meat has not generated further practical activities concerning the feasibility of this innovation. Quite the opposite is the case for the normative and social aspects of this innovation.

4 Envisioning the Ethicality of In vitro Meat

Reducing the number of animals required for the production of meat is one of the ethical advantages of IVM. In the relevant literature the vision can be found that one animal alone might be sufficient for meeting the worldwide meat demand (Bhat et al. 2015). Even if this may be an exaggeration, one could imagine that reducing the number of animals might make factory farming unnecessary, so that the few remaining animals could be kept under better conditions.

Our interviews with different stakeholders pointed out further ethical aspects: In vitro meat could be a transitional solution on the way to a world where animals have rights and are not exploited any more. Different from the innovators' argumentation, in vitro meat is understood as a means to another end: realising a society in which animal products are abandoned and humans and animals live as equals. This argument is reconstructed as the pragmatic argument: According to a representative of an animal rights organisation, the exploitation of animals, which basically includes animal husbandry and the killing of animals, is an outcome of speciesism and should therefore be abolished. That means a vegan or plant-based diet is the actual solution to animal exploitation and the current problems of meat consumption and production. Eating meat is an eating and cultural habit, so it would be better if people eat in vitro meat instead of conventional meat. In order to change these habits, conventional meat must be put into question. In vitro meat could raise the issue and therefore strongly support animal rights by weakening "the animal exploitation industry" (B78).

However, the prophesied "liberation of animals" is not (yet) possible, also because of the use of other animal products, especially foetal bovine serum. Other elements of the production process include animal products too, e.g. growth factors and the materials for frames. The innovators attempt, among others, to replace foetal bovine serum by alternatives (such as seaweed, yeast). A tissue engineer raised the question whether people would accept in vitro meat if the meat itself would be grown from cells, but the coating still would come from rat's tail (see Interview J). The biggest question, however, is as yet unanswered: given IVM, what will the future of animals look like? How will animals live?

In addition, an in vitro meat researcher stated that in vitro meat could also be a chance to regain respect for farmers by revaluating meat and agriculture. The loss of value and respect is attributed to intensive farming and excessive meat consumption. In vitro meat could support agro-ecological farming by being integrated into agro-ecological farming practices:

"I think the biggest thing cultured meat could do is to reshape the agricultural system. (…) Hopefully, revitalise agriculture." (I228 ff.)

The representative of an animal rights organisation also thinks that farmers are pushed economically and would act differently if they had the possibility to do so (B65). This line of thought is found in an interview with an in vitro meat researcher as well:

"[Farmers are] being pushed economically to farm against nature instead of with it. And we need to support those types of farmers and hope that cultured meat could help them become economically competitive with the large-scale ones that we're trying to get rid of." (I254 ff.)

The interviews with stakeholders showed that they believe the transition from the present to the future according to the vision might become a failure. Many stakeholders believe that society lacks awareness of the problems of meat production and consumption.

However, some of these interview partners did not agree with the basic assumption of the innovators: The conventional producer cooperative's representative and the representative of the food service company do not share the innovators' initial problem. They do not consider today's meat production a problem. In their view, 98% of livestock holdings, most of which are family-run businesses, are alright (K69). Meat consumption has already reached its peak, is slightly declining and will level off (K33, K285). High meat consumption is mainly regarded as a health problem (L109); but meat consumption as such is seen as part of a balanced diet (K36). So the problem to which the innovation of in vitro meat is presented as a solution is-in their view-non-existent. The basic premise of the innovators therefore seems to be rejected.

Some stakeholders accept that the innovators have strong arguments, in particular those belonging to animal rights organisations. Thus, in vitro meat is considered the quickest way of changing the consumption of meat; this illustrates the normative power of this way of arguing. However, this innovation is strongly rejected by supporters of alternative animal husbandry who are not convinced of the normative power of the innovators' arguments. Due to its function as a communication medium, the vision also allows for the food industry's position which considers the communication of the added value of in vitro meat essential for its successful marketing.

In general, a strong polarisation concerning the innovation of in vitro meat could be identified, both when it comes to feasibility and desirability.

Accordingly, in the interviews on the one hand the normative power of the vision of in vitro meat is reflected by the pragmatic argument in support of it. In vitro meat might be a tool for a "techno-moral change", "an opportunity for a change of thinking" (van der Weele and Driessen 2016). This argument defends in vitro meat for pragmatic reasons because one considers this innovation the driving force for a far-reaching cultural change. In vitro meat might thus be the "second-best" one could do for animals, which is why it should be supported (Stephens 2013). From these points of view it would be best if animals would no longer be exploited for the satisfaction of human needs. However, from the point of view of those arguing, this is not feasible for the time being.

"I think in vitro meat will be a transitional solution. Consumers have to ask themselves: 'Do I have to kill an animal to eat meat?' The answer is: 'No.' The function of in vitro meat will be to raise that question and to reduce consumption of conventional meat. People will then realise that plant-based alternatives are better than in vitro meat. (…) In vitro meat can be a ground-breaking innovation for the status of animals in society. It will not be the end but a major step in the right direction." (B196 ff.)

On the other hand, an opposite normative power is developing based on the arguments of those who do not imagine in vitro meat as the solution to current problems and consider ecological animal farming and a reduction of meat consumption an alternative:

"That's our approach: Mainly plant-based and back to the Sunday roast." (A33 f.)

"Less meat and more vegan and other food is the right global approach. It's reducible to that." (D251 f.)

From the organic producer association's point of view, the consumption of meat is declining, at least in Germany. So the reduction of meat consumption seems to be a viable solution for them (A67), and it seems not very likely that in vitro meat will catch on in society-mainly due to cultural and taste aspects (A77). In vitro meat is seen as artificial meat and not as the same thing as "real" meat. So according to NGOs and a politician, the best solution for current problems would be a reduction of meat consumption by half and organic farming with animals (A67, E278) or small-scale farming (D29) and *not* in vitro meat (cf. Klima-Allianz Deutschland 2016).

This is in immediate contradiction to the opinion of the innovator that the consumption of meat is increasing and the reduction (or termination) of meat consumption is unlikely-because people love eating meat (C204, C449). In vitro meat, which after all is considered to be the same as "real" meat, is therefore seen as the most realistic solution.

Thus, the innovators' argumentation was strongly questioned by several interviewees. Perceiving the same problems of today's meat consumption, they proposed a different and, in their view, fairly obvious solution approach. According to the representatives of the environmental organisation, the organic producer association and the politician, consumption has reached its zenith (e.g. D246). In the Western world we eat too much meat; so "there lies a great potential for change" (A25).

The proclaimed solution to the problems of today's meat consumption and production is reducing the consumption of animal products, a mainly plant-based diet and regionally adapted organic farming with animals. In the end, not only one solution will be realised: There will be a mix of different protein sources (C200, C409, C498, D260), like plants, insects (A354, A358, D23, D255), algae (J52) or food from fermenters (A290).

Furthermore, a representative of an environmental organisation said that in vitro meat production would accelerate the alienation of consumers and animal production. It would be another step towards industrialisation and therefore towards an alienation from actual 'agriculture', including animal husbandry. He actually fears the loss of agro-ecological farming (E28, E69, E74). In the worst case, in vitro meat could be, in the view of an organic producer association, counterproductive and could lead back to today's problems of high meat consumption: people might start consuming even more meat because it will have become ethically correct-there would be no need to change their behaviour (A43, A200, A205) or to think about consumption anymore (I172). So there is a shared concern of the NGOs and the in vitro meat researcher about losing respect for meat (and the animal behind it). But the in vitro meat researcher is more optimistic that in vitro meat could help with revaluating meat rather than contribute to devaluating it (I163).

Thus, there is a shared perception of current problems by the innovator, the in vitro meat researcher, NGOs, the politician and the organic producer association, but with different prioritisations: There needs to be a change of meat production which in most cases intends the elimination of intensive farming and its negative impacts on the environment (D110, E95, J291), animal welfare (A414, D110), climate change as well as concerns about resource efficiency (D99, D105).

5   Stakeholders' Positions on In vitro Meat in Germany

During the time of the VIF project, and in particular during the last phase, some stakeholders in Germany decided to take up a position on in vitro meat. We are not saying here that these statements are immediate consequences of our scientific project. First of all because our interviews, whose results have been published anonymised, do not necessarily refer to and match the stakeholders who decided to write statements on in vitro meat. Second, because these stakeholders did not explicitly quote our project in their statement. However, we believe that the scientific activities of our project which gained visibility in the media (being at the time the first and only project in Germany on in vitro meat) as well as the increasing attention of the media for this innovation in general pushed the debate forward. These statements and reports are concrete activities which were generated by the engagement with the innovators' vision. Therefore they can be interpreted as socio-epistemic practices, together with the already explained visions pointed out during our interviews.

In April 2016 the non-profit organisation *Sentience Politics* published a policy paper on cultured meat. In this paper the NGO follows the innovators' vision, describing meat production and consumption as environmentally and ethically apprehensive. *Sentience Politics* supports efforts to "[f]und and promote academic interest in cellular agriculture", "[i]ncrease public awareness about the benefits of cultured meat", and "[f]acilitate cultured meat development through policy changes" (Rorheim et al. 2016).

Also some animal welfare and animal rights organisations in Germany reported on in vitro meat and commented very positively on this innovation, providing information on in vitro meat and representing the innovators' visions.[3] Animal Equality picked up on the vision that "billions of animals could be spared a life of misery and a cruel death" and that the mass production of in vitro meat would reduce the emissions of greenhouse gases and the use of land, water and energy, but also described different substitutes for meat like plant-based alternatives.[4] Vier Pfoten introduced different companies working on in vitro meat, but

---

[3]https://www.vier-pfoten.de/kampagnen-themen/themen/ernaehrung/clean-meat, https://www.animalequality.de/neuigkeiten/9-25-2015-das-fleisch-der-zukunft.
[4]https://www.animalequality.de/neuigkeiten/9-25-2015-das-fleisch-der-zukunft.

used the term "clean meat" to depict that it is cruelty free—comparable to "clean energy". This term describes important aspects of the technology like "positive effects on the environment" and "absence of pathogenic agents and pharmaceutical residues". Referring to an interview with Mark Post, they consider its acceptance to be high.[5]

Gerald Wehde, spokesman of the ecological agriculture association Bioland, expressed rather sceptical appraisals[6] on the association's web page, which is made clear by the title in the first place: "No one needs meat from the lab!" He is "convinced that the public will not accept artificial meat" and argues that "80% of the Germans decline genetically engineered food and 70% decline cloning of livestock", raising the question how they would buy "genetically engineered meat, cloned burgers?" Consumers rejecting any kind of livestock farming for ethical reasons would certainly not eat artificial meat. Also "two-thirds of the global agricultural land are green land" which is used by livestock and could not be used for other purposes, and cows are needed not only for meat but also for milk production. Wehde refers to "regionally and culturally different eating traditions" and agricultural conditions, so there have to be different ways for future protein supply, for instance with insects. He sees the only solution in a global reduction of meat consumption and "ethically acceptable livestock farming". In the last days of writing this article, one big player of the German meat industry, the poultry farmer *Wiesenhof*, openly declared to subsidise the Israeli start-up *Supermeat* dedicated to the development of in vitro meat.[7] A few days later another important stakeholder in Germany took position: the German Farmers' Association expressed scepticism about the potential of in vitro meat.[8] Last but not least, Brandenburg's Farmers' Association positioned itself against this innovation and summoned the Federal Government to stop any investment therein.[9] We are expecting an exciting debate on this innovation in Germany in the time to come.

---

[5]https://www.vier-pfoten.de/kampagnen-themen/themen/ernaehrung/clean-meat.

[6]http://www.bioland.de/im-fokus/artikel/article/fleisch-aus-dem-labor-braucht-kein-mensch.html.

[7]https://www.abendblatt.de/wirtschaft/article213027865/Wiesenhof-setzt-auf-kuenstliches-Fleisch-aus-dem-Labor.html.

[8]https://www.presseportal.de/pm/58964/3832376.

[9]https://www.topagrar.com/news/Home-top-News-Bauernbund-gegen-Kunstfleisch-Konzerne-greifen-erneut-nach-Macht-bei-Lebensmitteln-8991760.html.

6 The Significance of the Results for Technology Assessment

According to the innovators' vision, in vitro meat presents itself as the interface between the present state and the future of food production. In vitro meat mobilises different interests in society and serves as a point of reference by bringing them together. Due to the strong ethical claim of being the solution to the problems for man, animal and the environment resulting from meat production and consumption, the vision of in vitro meat becomes connectable to the sustainability discourse and can be discussed as a serious option for food policy (Ferrari and Lösch 2017).

Interviewing experts and stakeholders has proven its worth when it comes to making competing approaches to a solution visible. Some experts pointed out different technical problems in the development of this innovation which are underestimated or not completely visible in the innovators' vision. The lack of studies on LCA as well as the need for more basic research, which are both at the different investors' core of interest, emerged very well through different interviews. Among the experts, the ones more close to the innovators' vision tend to describe these difficulties as reason for more research and as positive, whereas the more critical experts tend to dismiss the potential of in vitro meat. At present, there are still no new publications on LCA or on the biological challenges of this innovation. This engagement leads to the elaboration of important data for technology assessment, but did not stimulate any further interest in scientific publications on the problems of this innovation.

However, the picture looks different regarding the ethical and social aspect of in vitro meat. Our interviews pointed out that when experts and stakeholders were confronted with the innovators' vision, the focus shifted from the innovation of in vitro meat towards the non-technological issue of future food. Even if the interviewed stakeholders had different focuses of concerns (the market, animals or the environment), they all tended to put in vitro meat in a broader discourse on the future of our nutrition (and meat) and the role of agricultural practices as we know them now. Once in vitro meat gained more visibility in the media, some stakeholders in Germany (not necessarily those we interviewed, as already stated) even decided to report on in vitro meat or to make statements in favour of or against this innovation. The vision of the innovators is therefore taken up by the societal discourse while, vice versa, affecting it, even if in contrasting ways. On the part of some animal welfare and rights organisations in Germany, in vitro meat has been favourably received, since it might trigger a critical analysis of meat in society which might contribute to making the public more aware of the many bad consequences of today's meat production and consumption. On the other hand, stakeholders from the field of ecological agriculture and environment protection as well as politics in Germany show a negative attitude towards this vision, stating that ecological agriculture in combination with a reduction of meat consumption by one half will solve the problems. In ecological agriculture, they state, animals play a crucial and indispensable role for the natural cycle. The ambivalence of the different and even dichotomous perspectives on in vitro meat clearly expresses the challenge which the current way of meat production and consumption poses to humankind-and the difficulties of finding alternatives that will be accepted by the broader public.

All in all, the direct confrontation with the innovators' vision permitted to gain not only interesting insights into this innovation, but showed that this innovation does not come in a vacuum for different societal actors, but develops from the conception and values concerning food and agriculture. If in vitro meat would come, it would not simply substitute (some) meat products, but would be a trigger of experiences, feelings and values concerning the whole food complex. Dealing with the visions of in vitro meat as socio-epistemic practices in the context of vision assessment has demonstrated how, by way of this vision, meanings are generated, activities are oriented and new social orders (e.g. new alliances between social groups) are formed. Thus, the analysis of technological innovations by help of the vision assessment of TA contributes considerably to evaluating their consequences for society and research, and contributes to identifying and assessing options for a promising, sustainable development and to making them part of the societal and political discourse.

References

Albritton, R. (2013). Between obesity and hunger: The capitalist food industry. In C. Counihan & P. van Esterik (Eds.), *Food and culture. A reader* (pp. 342–352). New York: Routledge.

Alexander, P., Brown, C., Arneth, A., Dias, C., Finnigan, J., Moran, D., et al. (2017). Could consumption of insects, cultured meat or imitation meat reduce global agricultural land use? *Global Food Security, 15,* 22–32.

Animal Equality. (2015). Das Fleisch der Zukunft. https://www.animalequality.de/neuigkeiten/9-25-2015-das-fleisch-der-zukunft. Accessed 8 May 2017.

Bhat, Z., & Bhat, H. (2010). Tissue engineered meat-Future meat. *Journal of Stored Products and Postharvest Research, 2*(1), 1–10.

Bhat, Z., Kumar, S., & Bhat, H. (2015). In vitro meat production. Challenges and benefits over conventional meat production. *Journal of Integrative Agriculture, 14*(2), 241–248.

Bioland. (2016). Fleisch aus dem Labor braucht kein Mensch! https://www.bioland.de/im-fokus/artikel/article/fleisch-aus-dem-labor-braucht-kein-mensch.html. Accessed 23 Feb 2018.

Churchill, W. (1931). Fifty years hence. *Maclean's Magazine.* http://rolandanderson.se/ Winston_Churchill/Fifty_Years_Hence.php.

Cultured Beef. (2017). Cultured meat. https://culturedbeef.org. Accessed 12 Apr 2017.

Cultured Beef. (2017). What is cultured beef? https://culturedbeef.org/what-cultured-meat. Accessed 12 Apr 2017.

Ferrari, A. (2016). Envisioning the futures of animals through in vitro meat. In I. Olsson, S. Araújo, & M. Vieira (Eds.), *Food futures: Ethics, science and culture* (pp. 265–270). Wageningen: Wageningen Academic Publishers.

Ferrari, A., & Lösch, A. (2017). How smart grid meets in vitro meat: On visions as socio-epistemic practices. *Nanoethics, 11,* 75–91.

Grin, J., & Grunwald, A. (Eds.). (2000). *Vision assessment. Shaping technology in 21st century society. towards a repertoire for technology assessment.* Berlin: Springer.

Grunwald, A. (2004). Vision assessment as a new element of the FTA toolbox. EU-US seminar: New technology of foresight, forecasting and assessment methods-Seville 13–14, May 2004, 53–67.

Grunwald, A. (2012). Synthetische Biologie als Naturwissenschaft mit technischer Ausrichtung. Plädoyer für eine "Hermeneutische Technikfolgenabschätzung". *Technikfolgenabschätzung-Theorie und Praxis, 21*(2), 10–15.

Grunwald, A. (2015). Die hermeneutische Erweiterung der Technikfolgenabschätzung. *Technikfolgenabschätzung-Theorie und Praxis, 24*(2), 65–70.

Heinrich-Böll-Stiftung & BUND (Ed.). (2016). *Fleischatlas 2016-Deutschland regional. Daten und Fakten über Tiere als Nahrungsmittel.* Berlin: Heinrich-Böll-Stiftung.

Institute for Technology Assessment and Systems Analysis (ITAS), Karlsruhe Institute of Technology (KIT). (2014). Project visions as socio-epistemic practices-Theoretical foundation and practical application of vision assessment in technology assessment. Karlsruhe: ITAS/KIT. https://www.itas.kit.edu/projekte_loes14_luv.php. Accessed 6 May 2017.

Klima-Allianz Deutschland (Ed.). (2016). *Klimaschutzplan 2050 der deutschen Zivilgesellschaft.* Berlin: Klima-Allianz Deutschland.

Mattick, C., Landis, A., Allenby, B., & Genovese, N. (2015). Anticipatory life cycle analysis of in vitro biomass cultivation for cultured meat production in the United States. *Environmental Science and Technology, 49*(19), 11941–11949.

Modern Meadow. (2016). Modern meadow. http://www.modernmeadow.com.

New Harvest. (2017). Mission and vision. http://www.new-harvest.org/about. Accessed 2 May 2017.

Post, M. (2014a). An alternative animal protein source: Cultured beef. *Annals of the New York Academy of Science, 1328,* 29–33.

Post, M. (2014b). Cultured beef. Medical technology to produce food. *Journal of the Science of Food and Agriculture, 94*(6), 1039–1041.

Rorheim, A., Mannino, A., Baumann, T., & Caviola, L. (2016). *Cultured meat: A pragmatic solution to the problems posed by industrial animal farming.* Switzerland: Sentience Politics.

Schaefer, G., & Savulescu, J. (2014). The ethics of producing in vitro meat. *Journal of Applied Philosophy, 31*(2), 188–202.

Stephens, N. (2013). Growing meat in laboratories-the promise, ontology, and ethical boundary-work of using muscle cells to make food. *Configurations, 21*(2), 159–181.

Top Agrar Online. (2018). Bauernverbund gegen Kunstfleisch: Konzerne greifen erneut nach Macht bei Lebensmitteln. https://www.topagrar.com/management-und-politik/news/bauernbund-gegen-kunstfleisch-konzerne-greifen-erneut-nach-macht-bei-lebensmitteln-9412294.html. Accessed 19 Feb 2018.

Tuomisto, H., & Teixeira de Mattos, M. (2011). Environmental impacts of cultured meat production. *Environmmental Science & Technology, 45*(14), 6117–6123.

Van der Weele, C., & Driessen, C. (2016). In vitro meat is a chance to rethink. In N. Stephens, C. Kramer, Z. Denfeld, & R. Strand (Eds.), *What is in vitro meat? Food phreaking issue 02* (pp. 57–59). Dublin: The Center for Genomic Gastronomy.

Vier Pfoten Deutschland. (2017). Clean meat. https://www.vier-pfoten.de/kampagnen-themen/themen/ernaehrung/clean-meat. Accessed 18 Apr 2017.

Wellesley, L., Happer, C., & Froggatt, A. (2015). *Changing climate, changing diets. Pathways to lower meat consumption.* London: Chatham House.

Interviews

The expert interviews are tagged with capital letters from A-L, the subsequent number refers to the line number of the interview transcript.

Interview A: Representative of an organic producer association, conducted: 13.6.2016

Interview B: Representative of an animal rights organisation, conducted: 15.6.2016

Interview C: Innovator and researcher, conducted: 15.6.2016

Interview D: Politician, conducted: 16.6.2016

Interview E: Representative of an environmental organisation, conducted: 22.6.2016

Interview I: In vitro meat researcher, conducted: 28.6.2016

Interview J: Tissue engineer, conducted: 30.6.2016

Interview K: Representative of a conventional producer cooperative, conducted: 13.07.2016

Interview L: Representative of a food service company, conducted: 19.07.2016

Inge Böhm worked at the Institute for Technology Assessment and Systems Analysis at the Karlsruhe Institute of Technology on a project on "Visions of in vitro meat". In the project, ethical and societal aspects of in vitro meat were analysed.

Silvia Woll is a philosopher and researcher at the Institute for Technology Assessment and Systems Analysis at the Karlsruhe Institute of Technology. Her work focuses on new and emerging technologies and technology ethics.

Arianna Ferrari (Ph. D.) is philosopher and head of the Department Strategy and Content at Futurium gGmbH, a new space in Berlin, which works as astage, museum, laboratory and forum of the future (www.futurium.de).

The Naming of the Beast. Scrutinizing Concepts of Technology Rich Futures

Jan Nolin, Gustaf Nelhans and Nasrine Olson

Abstract

Constructive Technology Assessment (CTA) aims to involve a multitude of stakeholders in discussions involving technology. However, all technological assessment is mediated through concepts. Development of concepts regarding technology rich futures has been lively but confusing for many stakeholders. Utilizing bibliographic data from Web of Science, this article overviews and discusses ten influential concepts that have played an important role during the 2000s. It is argued that the lively and dynamic concept production leads to numerous problems for a CTA approach.

1 Introduction

Technology is always understood through concepts. In a sense, original concepts need to be in place before the construction of innovative technology. Otherwise, value, use, impact and function of devices cannot be promoted, planned and funded. This coproduction of technology and concepts is also present when

J. Nolin (✉) · G. Nelhans · N. Olson
University of Borås, Swedish School of Library and Information Science,
Borås, Sweden
e-mail: jan.nolin@hb.se

G. Nelhans
e-mail: gustaf.nelhans@hb.se

N. Olson
e-mail: nasrine.olson@hb.se

© Springer Fachmedien Wiesbaden GmbH, part of Springer Nature 2019   231
A. Lösch et al. (eds.), *Socio-Technical Futures Shaping the Present*,
Technikzukünfte, Wissenschaft und Gesellschaft / Futures of Technology,
Science and Society, https://doi.org/10.1007/978-3-658-27155-8_11

discussing the future. This article presents 10 concepts used to talk about the future of humans, their societies and technology. In addition, the article discusses particular problems connected to the various conceptual tools that are available for us today to talk about tomorrow.

Developed economies of the late 2010s are already highly technologically dependent. However, much more is on its way. Internet-based technologies are evolving rapidly as everything from coffee machines to light bulbs and furniture are given IP numbers. An avalanche of innovations with mounting economic and social impacts is pushed on marketplaces globally. Such fundamental changes in the material aspects of everyday life raise numerous challenges for a variety of stakeholders such as social scientists, philosophers, policymakers, journalists etc. What will all these technological innovations mean for our society and us as human beings?

However, stakeholders have had difficulty in "playing catch up" to a quickly evolving spectrum of new technology. Put bluntly, scrutiny by stakeholders becomes fragmented, attending to diverse concepts and visions. In this article we argue that a substantial aspect of this fragmentation stems from engaged stakeholders easily becoming derailed when attempting to understand visions through a variety of concepts. An obvious problem when discussing concepts that are used to talk about the future is that there is a need for a neutral concept that does not depend on the other concepts analyzed. We have chosen "technology rich futures" as such a neutral concept. Consequently, it refers broadly to visions in which societal dependence on technology has been escalated considerably beyond current situation.

We will discuss bibliographic data connected to 10 concepts that articulate variations of a technologically rich future. We will focus on three dimensions of "conceptual messiness" within ICT research:

- *diversity*: there are many concepts,
- *domains*: technology can be talked about in a variety of ways and
- *lifecycle*: concepts come and go over the span of a few years.

The article starts with a presentation of theoretical departure and some remarks on method. Thereafter, the 10 different concepts are reviewed, illustrating conceptual diversity. This leads to a discussion on domains and lifecycles.

2   Theoretical Departure

Traditional Technology Assessment (TA) approaches innovations years after implementation, supplying an autopsy of transformative processes. Given the quickening pace of technological innovation, such a measured approach appears increasingly problematic. One strategy that has been touted as a way for dealing with such problems is Constructive Technology Assessment (CTA) which is an attempt to increase the relevance of traditional TA, involving a socio-technical critique of technological developments (Schot 1992; Schot and Rip 1996). While TA dealt with retrospective assessments of mature technology, CTA attempts a proactive strategic intervention with reflexive and anticipatory orientation. CTA is concerned with "democratic reflexive discourse" (Genus 2006), promoting integration of societal concerns in technological change.

CTA pushes for engagement among a broad range of actors including various groups within the general public, politicians, developers, scholars (especially from the field of STS), organizations, societal groups and other actors before dramatic technological change takes place. The insights gained in these discussions are to inform technology design, development, management and decision making. In recent years the CTA approach has been applied in medical research (e.g. Douma et al. 2007), nano-technology (e.g. Rip 2008) and ICT planning in developing countries (e.g. Moens et al. 2010).

Historically, milestones in the evolution of communication and Information and Communication Technology (ICT) (e.g. invention of writing systems, production of paper, and development of printing technologies) have accompanied major societal changes. This is sometimes described as coproduction of technology and society (Guston and Sarewitz 2002). *However, the coproduction of technology and society should also be understood in the context of coproduction of innovative concepts.* In other words, the broad range of stakeholders that are to be involved with the CTA approach needs to engage with appropriate concepts in order to deliberate the technology of the future. Furthermore, all assessment of technology is mediated by clusters of terms specifically associated with various artefacts. This issue has not been clearly elaborated on within earlier research on CTA.

We will utilize the notion of *conceptual domain*, which has mostly been used in research on metaphors (Lakoff and Johnson 2008). According to this tradition, concepts can be seen as situated in different domains together with associated concepts. Metaphors are used to make abstract phenomena, such as technology rich futures, more concrete. Resources are therefore taken from one conceptual domain to attempt communication of phenomenon otherwise situated within another domain. For the purpose of this article, the domain that is to be explained (with the help of other domains) is *technology*. After having reviewed the 10 concepts in focus, we will then identify nine different conceptual domains that are used to talk about technology rich futures.

We will also utilize the notion of *conceptual landscape* (van Raan 2017) as such become possible through the visualization device Vosviewer (van Eck and Waltman 2010). We will review concepts found in titles and abstracts. The application will utilize different colors to represent average publication years for texts. Large nodes will indicate frequent usage of terms. Proximity between nodes will illustrate the degree of connectedness across texts. With this application it will be possible for us to identify the presence and emphasis of various conceptual domains as they manifest themselves in the visualization regarding 2008–2016 (see Fig. 2). Building on that, we can see shifts in emphasis over time regarding individual domains. In addition, we will be able to identify heavily emphasized domains as well as those less visible.

Reviewing the dynamics of different conceptual domains enables us to draw conclusions regarding the conceptual challenges facing stakeholders.

3 Method

A wide array of concepts is regularly developed for R&D, marketing and distribution of new technology. Many more emerge as people use devices in their everyday life. However, the concepts to be used in visionary attempts to describe the role of technology in the future societies constitute unique challenges to our language. Most stakeholders lack the necessary insights into emerging technology to suggest appropriate concepts.

The basic methodological idea underpinning this article is that researchers are either inventors or early adopters of new concepts aimed at describing technology rich future societies. A review of available scholarly texts registered within the aggregated database of *Web of Science* should therefore provide valid insights into the emergence and development of such concepts. This approach, of using scholarly publications as a resource for investigating concepts, can be seen as an offshoot to bibliometrics. Traditionally, the main focus of bibliometrics has been the study of citations, bibliographic coupling and co-citation (Kessler 1963; Small 1973; Garfield 1979). However, bibliographic data can also be useful for the investigation of emerging concepts.

In the following we extend an earlier study (Olson et al. 2015) where we attempted to take a full inventory of broad concepts used to talk about a technology rich future. In that review we avoided concepts that point to a more concrete cluster of technologies (e.g. cloud computing), technological methods (e.g. big data), industrial shift (e.g. the fourth industrial revolution) or notions of economic development through technology (e.g. data-driven innovation). Such concepts would, no doubt, be interesting to include in discussions of the challenges of attaching stakeholders to technological visions. However, we have in the current text attempted to demarcate our material and limit ourselves to broad, "umbrella like" concepts concerned with technology rich futures.

14 concepts emerged from an extensive review of vision oriented texts in ICT research. Thereafter, these were tracked within Web of Science, identifying concepts of most frequent usage within scholarly literature. This led to a focus on the 10 most active concepts on technology rich futures in Web of Science (including SCI-EXPANDED, SSCI, CPCI-SSH, A&HCI, CPCI-S, and CPCI-SSH) published 1971–2016. This covered not only the technological and natural sciences, but also the humanities and social sciences. The aim was to identify both diversity of concepts and their longitudinal development over time as they appeared in the literature of all fields of research. The data on longitudinal development data will be presented in Fig. 1. We also collected the various terms used in titles and/or abstracts of the 37,384 publications. This data set, combined with the longitudinal data 2008–2016, will be presented in Fig. 2.

We have found that bibliographic data is a powerful resource in studying the dynamics of concepts coproduced with technological development. However, there are also some limits with this approach which should be noted. Using data from an exclusive database such as Web of Science means that only material that is published in carefully vetted journals and conference series are indexed. This is not a substantial problem for the current study as the intent is not to write the history of each of the distinct concept in focus. We are more concerned with trends in the bulk of the literature produced, to identify the evolution of themes that pop up and the life cycles of each of the terms' use. Therefore, a certain absence of specific documents-even of important ones-can be disregarded. To a certain extent, it is also possible to check what material has not been covered. Works that do not use the specific terms within their titles, abstracts or keywords, but still are relevant could be identified based on having been cited among the covered material retrieved from the database.

Indeed, a total of 508,611 documents are identified as cited from the total of 37,384 documents in the data set. Reviewing the cited set, we find that 5349 documents have been cited at least ten times. Among the cited material outside the data set we find journal articles that are not published in indexed journals as well as popular articles and books (including monographs, text books, edited works). While it would have been interesting to include all of this material, it is practically impossible as not everything is digitally available. Furthermore, that something has been cited in a text does not necessarily mean that it is relevant for the case at hand.

Another issue concerns the time lag between submitting a manuscript and the actual publication of the paper. This means that some articles may be published one

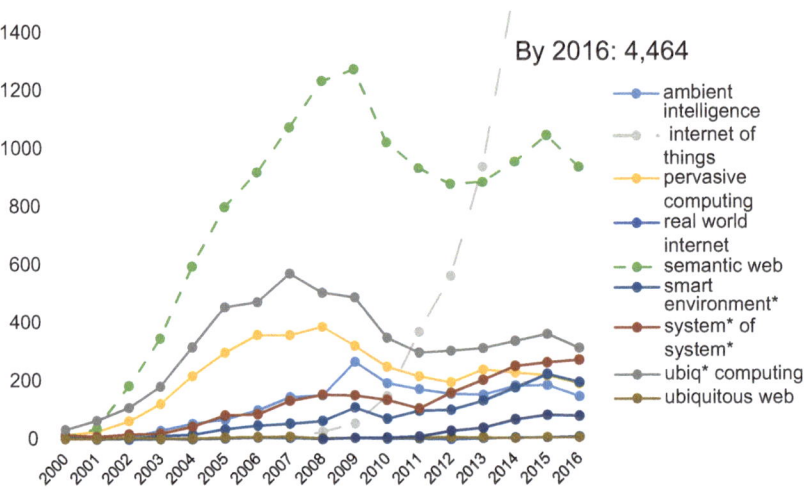

Fig. 1 The longitudinal development of 10 concepts describing a technology rich future measured according to mentions in number of texts within Web of Science 2000–2016

or two years after having been written. Again, since we are looking at aggregated data rather than individual historical events, there is a similar lag implied in all publication and therefore changes can be expected to be "smoothed out" rather than leading to sharp breaks between different periods. Having these difficulties in mind, we are still confident that the current approach allows for the best available method for overviewing the development of concepts describing a technology rich future.

4   Diversity of Concepts

In the following, we will briefly review the 10 concepts that have been used within ICT research to talk broadly about technology rich future societies.

Figure 1 visualizes the development of the 10 concepts that will be discussed below within the space of this millennium. As is obvious, there are two concepts that can be described as "outliers": semantic web and Internet of things. The former peaks in 2008 while the latter expands in usage dramatically a few years later. It should be noted that Web of Science has a yearly expansion of approximately 5%. This leads to concepts on track of being used more frequently each year. Therefore, a distinct decrease in usage, notable for several concepts following a peak, signals a dramatic shift in scholarly discussions. In the following, we will present each of these concepts in the order of chronological emergence.

4.1   System of Systems

This term dates back to the early 1970s, mostly within literature relating to the US army effort of revolutionizing logistics in combat situations through what was called *Army After Next* (AAN) (e.g. Badger et al. 2012; Kelley and Pei 1999; Knichel 2010; Marquet and Ratches 1998). The system of systems was poised to exploit sensor and satellite data, processing and disseminating information in the context of information warfare. System of systems was initially situated within a military mindset of command and control, enabling highly informed key decisions by commanding officers at the supra-system level and allowing such commands to infuse all other systems. The concept became, with time, somewhat divorced from its military origins. In recent years it has become used as a way to talk about the linkages of systems and devices. The term occurred in 2214 documents in our study. As illustrated in Fig. 1, the frequency of use has in recent years been increasing.

4.2   Ubiquitous Computing

This concept came out of the Xerox Palo Alto Research Center in the 1980s, referring to a society in which human computer interaction would become seamlessly integrated into everyday life (Weiser 1991, 1993a, b, 1995; Weiser et al. 1999). A formative notion is that humans should not adapt their activity to that of machines. Rather, the technology should be humming in the background, dynamically creating the appropriate environment for human activities (e.g. Weiser 1993a, p. 76; York and Pendharkar 2004). The term occurred 5626 times in our material. However, it had peaked by the turn of the first decade of the 2000s although used regularly thereafter.

4.3   Pervasive Computing

This term appears to have moved from AT&T and IBM into the research literature (Satyanarayanan 2001). Orwat et al. (2008, p. 2) find pervasive computing to be "loosely associated with the further spreading of miniaturized mobile or embedded information and communication technologies (ICT) with some degree of 'intelligence', network connectivity, and advanced user interface". With time, the concept grew increasingly close to that of ubiquitous computing. As with that concept, visions put forth by Mark Weiser remained central (e.g. Saha and Mukherjee 2003; Satyanarayanan 2001; Yachir et al. 2012). Despite the close connection, both terms have developed separate conferences, journals and communities. The term occurred 3779 times in our material and rose in popularity from 2001, peaking seven years later, in 2008. However, it remained an important concept in our material with some 200 articles published yearly.

4.4 Semantic Web

This concept expands on the idea of World Wide Web created by Tim Berners-Lee (later described in e.g. Berners-Lee et al. 2001). Semantic Web deals with web-based information, information access, knowledge representation, and semantic codes appropriate for technology intervention. The main focus is shifting the character of content from something to be read by humans toward advanced processing by computers. Semantic agents and similar assistive technology would serve humans by processing, aggregating, recommending and networking information. The concept was developed during a period when ubiquitous computing was widely discussed in ICT research. Utilizing the notion of semantic agent, Berners-Lee and colleagues envisioned smart technology in the form of "personal assistants" that processed tasks, explicitly directed by users. This was in contrast to the approach taken by Weiser (1993a, p. 76) who envisioned IT resources humming unobtrusively in the background. The semantic web was the most frequently used concepts in our material, with 13,314 articles involving 33% of the data. The concept rose dramatically in popularity from 2001 and peaked in 2009, the year in which almost 1,300 articles featured this concept. By 2013, the yearly production of articles focusing this concept was below 900.

4.5 Ubiquitous Web

This concept was introduced in the late 1990s (e.g. Liu et al. 1998; Logenthiran et al. 1998) as a way to discuss the explosive development of web access. However, it soon came to refer to an integration of telephone and web-based services (e.g. Huang et al. 2000) as well as context aware computing and personalization (e.g. Finkelstein et al. 2002). The concept came to focus mobility as well as information (rather than data). Billsus et al. (2002, p. 34) discussed "any information at any time from any location". Core is the idea that there is a problem in the way that available information is so plentiful and the matching interfaces of mobile device are unequal to the task. If the full range of information were to be presented on mobile devices, it would burden the user rather than facilitate information access. Here, adaptive interfaces (that would learn from users' past behavior and interests) and personalization are presented as solutions for mobile access to web-based information (Billsus et al. 2002). That is, only the most relevant information (based on user's previous interests) is to be presented within the limited space available on mobile interfaces. Ubiquitous Web is somewhat narrower than the concept of ubiquitous computing in that it places focus mainly on mobile, web-based information use, rather than a wider range of technologies. However, a more important difference is the introduction of directed advertising and commercialisation of web-based information with the assumption that direct advertising "benefits considerably" from personalization (e.g. Billsus et al. 2002, p. 37). Here, a different sense of the user emerges, where the technology could benefit commercial corporations as the user, rather than the individuals who seek and use web-based information. This term has only occurred 111 times in our data. Nonetheless, it has remained valid all through the 2000s.

4.6 Internet of Things

This concept was probably coined in 1999 (Ashton 2009), initially intended as a way to talk about RFID technology and Internet connectivity. This basic meaning has broadened substantially over the years. Sundmaeker et al. (2010, p. 41) stated that the "Internet of things (IoT) is a dynamic global network infrastructure with self-configuring capabilities based on standard and interoperable communication protocols where physical and virtual 'things' have identities, physical attributes, and virtual personalities and use intelligent interfaces, and are seamlessly integrated into the information network". The term grew slowly in use from 2001 but didn't really take off until 2011 when close to 300 articles were published. By 2016 4464 articles included this concept. Altogether in our data the concept occurred in 11,930 articles.

4.7 Ambient Intelligence

This term emerged from a team at Palo Alto Ventures comprising of Eli Zelkha, Brian Epstein, Simon Birrell, and Clark Dodsworth and first presented by their colleague Roel Pieper (Zelkha and Epstein 1998) at the Digital Living Room

Conference organized by Philips Research. This concept was later used by the Information Society Technologies Advisory Group, an entity that provides independent advice to the European Commission, in their discussions of visions for the future (e.g. Ducatel et al. 2001). Augusto and Aghajan (2009, p. 1) clarified that the idea is that "by enriching an environment with technology (sensors, processors, actuators, information terminals, and other devices interconnected through a network), a system can be built such that based on the real-time information gathered and the historical data accumulated, decisions can be taken to benefit the users of that environment". As with ubiquitous computing, technology will adapt to and anticipate human activities. However, it can be said to be differentiated from ubiquitous computing in the heavy emphasis on smart materials and other innovations that "truly integrate with life, in which our environments become wiser, more comfortable, and more compelling" (Zelkha and Epstein 1998). At the core of this vision is the idea of a "life-enhancing" environment that anticipates and fulfils "our" needs, without mediation, in order to make us more comfortable. This vision extends the range of technologies that are considered to go beyond computing. This term was featured in 2077 articles. It rose quickly in popularity from 2001, peaking in 2009 with more than 250 articles. Nonetheless, the concept remained relevant in the 2010s with slightly less than 200 articles highlighting ambient intelligence yearly.

4.8   Smart Environment(s)

This term is closely related to ambient intelligence. According to Augusto and Aghajan (2009, p. 1) it "refers to environments that sense, perceive, interpret, project, react to, and anticipate the events of interest and offer services to users accordingly". Furthermore, Augusto and Aghajan (2009, p. 2) uses the term "to emphasize the physical infrastructure (sensors, actuators and networks) that supports the system", involving decision-making through user centric data extraction. For Cook and Das (2005, p. 3) smart environments are "able to acquire and apply knowledge about an environment and also to adapt to its inhabitants in order to improve their experience in that environment". Making some comparative notes about similar concepts, Augusto and Aghajan (2009) argued an emphasis on intelligent interconnection of resources. It should also be noted that numerous similar concepts have been pursued such as *Intelligent Environments* (Brumitt et al. 2000) or *Responsive Environments* (Krueger 1977), which are often used synonymously. We found the term featured in 1433 articles. It is notable as a concept that in our material has not peaked. Rather, it is persistently increasing, although slightly, on yearly basis, being the subject of about 200 articles in 2016. This increase might also reflect the 5% yearly expansion of the Web of Science database.

4.9 Web of Things

This concept was inspired by both the Internet of things and the semantic web. Building on the further notion of Web 2.0 (O'Reilly 2005), it was suggested that the web should be expanded to integrate physical objects (Guinard and Trifa 2009). As the focus was placed on a web-based environment, the Web of things appeared to be a more specialized concept compared to the Internet of things. Nonetheless, it is closely tied to the World Wide Web expansion projects of *linked data* and *Web of data* (Bizer et al. 2008). The concept was included in 339 articles but remained steadily increasing in our material.

4.10 Real World Internet

This term was developed within projects financed by the European framework program 7 and in connection with discussions on the Internet of things. A focus emerged regarding the integration of on-line and off-line technology, integrating the "real world" into the internet. In addition, there was an emphasis on a shift from human to machine centric information exchange. Accordingly, concepts such as "identity", was extended to include objects. Devices are therefore both "consumers and producers of content" (Gluhak et al. 2009). Real World Internet was designed to be broader than the Internet of things as the former included visions regarding physical-virtual interconnection. It was also positioned as an umbrella concept, including notions such as *Internet of services* and *Internet of energy*. It is included here as an example of a concept that so far has not been given much traction, although probably not due to a lack of relevance. Rather, the concept Internet of things appears to have been broadened to include the various dimensions outlined in Real World Internet. It is only included in 69 articles in our material.

4.11 Problems of Diversity

In order to discuss future technology we need to access concepts that describe artefacts before they become constructed. This is particularly the case when articulating future technology dressed up as revolutionary. In the case of technology rich futures, we have found a wealth of conceptual innovators competing for the right to name that which is becoming. This intense competition in naming the future constitutes a persistent difficulty in any CTA inspired attempt at proactive discussions of technology rich futures.

A main argument in the current text is that the diversity illustrated by the 10 concepts discussed above constitutes an obstacle for stakeholders. Typically, stakeholders will be confronted with one concept at a time. Attempting to find more information, they will search for other texts connected to a keyword such as smart environment. In Olson et al. (2015) we noted that discussions on different concepts and scholarly works were seldom connected with each other.

We would like to emphasize that this rich diversity of concepts denoting technology rich future leads to two separate problems. These can be called the problems of *fragmentation* and *pioneering*.

The fragmentation problem emerges as different critical investigations become attached to various concepts. Therefore, discussions on individual concepts may be relevant for several concepts but, nevertheless, receive no wider reception. For instance, a concerned policymaker may be attempting to find critical scrutiny of the Internet of things, not realizing that there have been valuable investigations done targeting ubiquitous computing or pervasive computing. In other words, stakeholder based knowledge generated in connection with a particular concept may to a lesser degree be exploited and re-used.

The pioneering problem is a further consequence of the fragmentation problem, identifying a process in which non-technological stakeholder discussions tend to be restarted, ignorant about each other. For instance, a social scientist investigating *system of systems* might find a lack of earlier research from a societal perspective. Naturally, the researcher becomes inclined to take on the role of a pioneer. However, publications attached to the *semantic web* might constitute precisely the kind of previous research needed. Given the pioneering problem concerned stakeholders become more likely to start from scratch, rather than build upon the work of others connected to other concepts.

5   Conceptual Domains

With our qualitative review of the key texts surrounding the 10 concepts, it became possible to identify nine different conceptual domains, i.e. ways of talking about technology. Order of presentation is, in falling order, based on apparent importance in key texts.

The first conceptual domain is *smart*. Technology can be talked about as intelligent; it can learn, have knowledge and can acquire language skills. It should be noted that "smart" and associated concepts are taken from the domain of human intelligence. In the context of the material at hand, "smart" constitutes a metaphorical resource for identifying similar traits within technology. The smart domain is present in most of the concepts.

The second domain is *environment*. Technology can be talked about as marking out various types of space such as rooms, buildings and cities. The environment domain is particularly highlighted in ubiquitous computing, pervasive computing, ubiquitous web, smart environment and Real World Internet.

The third domain is *tool*. Technology can be talked about in terms of devices or artifacts for human-computer interaction. The tool domain is an obvious resource for talking about technology and it has a large presence. It is particularly emphasized in Internet of things, ambient intelligence and web of things.

The fourth domain is *network*. Technology can be talked about as being connected in chains, distributed and held together through a wealth of nodes. The network domain is an obvious resource for talking about technology and particularly present in system of systems, semantic web, ubiquitous web, Internet of things and web of things.

The fifth domain is *control*. Technology can be talked about as supplying power, to command, construct hierarchies and review procedures. The control domain is most heavily emphasized in the military ideas of system of systems. However, it is also significant in the Internet of things and Real World Internet.

The sixth domain is *user*. Technology can be talked about as being connected to humans and being of use for the activities that humans do. User driven personalization is at the core of the ubiquitous web. In addition, the user domain is visible in discussions connected to the system of systems, involving decision-making of central people in an organization. There is also a visible user perspective in ubiquitous computing, pervasive computing and ambient intelligence.

The seventh domain is *movement*. Technology can be talked about as enabling mobility. The movement domain is often used in connection with the environment domain, for talking about how technology is used when people move from one place to another. It appears crucial within the system of systems, ubiquitous computing, pervasive computing, ubiquitous web and ambient intelligence.

The eighth domain is *society*. Technology can be talked about as benefiting social groups and organizations. Similarly, technology can be discussed as being governed by states, authorities and policies. The society domain is somewhat present in ubiquitous computing and Real World Internet.

The ninth domain is *business*. Technology can be talked about as enabling innovation, entrepreneurship, efficiency and profit. The business domain has a certain presence within the ubiquitous web as benefits of direct advertising are highly touted.

In the previous section we identified a fragmentation problem of various stakeholders attaching themselves to different concepts. In this section, we similarly identify a *domain fragmentation* as different concepts from various domains can be used as resource for talking about technology in separate ways. Discussions of technology rich futures are fundamentally open-ended and technology can be embraced in a wide variety of ways. It matters if a concept is used which highlights the tool domain rather than the user domain, i.e. if the focus is on things rather than people.

6   Lifecycle of Concepts

Concepts denoting technology rich futures are not only plentiful and diverse; they also seem to move through relatively short lifecycles, as illustrated in Fig. 1. The lifecycle pattern of technology driven concepts coming and going can be challenging for stakeholders. While a community of computer scientists can orchestrate a move from one concept to another, other stakeholders may repeatedly be frustrated as they engage with concepts at a late stage of the cycle. At that time, computer scientists may already be losing interest, moving on to another term.

Our discussion of lifecycles can be linked to the notion of a hype cycle of emerging technologies as introduced by Gartner as early as 1995. Although investigations inspired by Gartner such as Feen and LeHong (2011) focus emerging technologies, Gartner's research seems equally applicable to concepts used to talk about technology such as big data and Internet of things. The argument is that a "technology trigger" leads to unreasonably high expectations. As enthusiasm for the technology spirals to a "peak of inflated expectations", a backlash emerges in the form of "negative hype", pinpointing weaknesses of the particular technology. This then leads to a "trough of disillusionment" and the technology then being positioned on a more reasonable track of expectations.

Five of the 10 concepts discussed in this text went through a distinct development of quickly trending, peaking and then reaching a kind of steady-state at a lower level of measured scholarly publication output. Building on our data and the numerous concepts, it could be argued that the "positive hype" impacts ICT researchers and other stakeholders differently. Compared to the visualizations put forward by Gartner the positive hype in our data is a much slower process, taking some 7–8 years to evolve into a peak. Non-technological stakeholders may need that time to be alerted to the new concept. However, at that time, ICT researchers may already be hyping a new concept.

6.1   From Semantic Web to Internet of Things

In Olson et al. (2015) we concluded our data collection with 2013. At that time, we identified one exception to the lifecycle pattern: the Internet of things. It had modest growth until 2008 when substantial expansion started, immediately passing the semantic web as the most frequent concept. Also, the semantic web peaked in 2008, the same year when the Internet of things started growing. This growth was still evident at the close of 2013.

In the update to our material, visualized in Fig. 1, this tendency has been dramatically extended. Given that, it is quite possible that the pattern of lifecycles involving competing and overlapping concepts emerging, then faltering, may be, at least temporarily, put at hold. If so, the concept of the Internet of things might emerge as a robust and resilient concept that stakeholders can commit to. In Fig. 2 we will take a closer look at the most frequently used words 2008–2016.

6.2   From Abstract to Concrete Concepts

Figure 2 illustrates a conceptual landscape of the longitudinal development of the 10 concepts 2008–2016, also including other terms frequently found in titles and abstracts. Large nodes indicate frequent usage of terms. Proximity between nodes illustrates the degree of connectedness across texts. Colder colors are reserved for usage in older publications (left side) while warmer colors signify recent publications (right side).

Figure 2 reveals the fundamental dominance of these two concepts with the semantic web as the most highly used in 2008. Although that concept remained highly mentioned within publications in the years to come, it was surpassed by Internet of things in 2013. Figure 2 therefore shows on the left side the main concepts clustered around the semantic web. These are primarily theoretical in character: *language, knowledge, ontology* and *web*. On the right side, with concepts associated with the most recent publications, we find the Internet of things as the clearly dominant concept. It is easy to observe more concrete concepts such as *physical device, smart building, power consumption, monitoring system.*

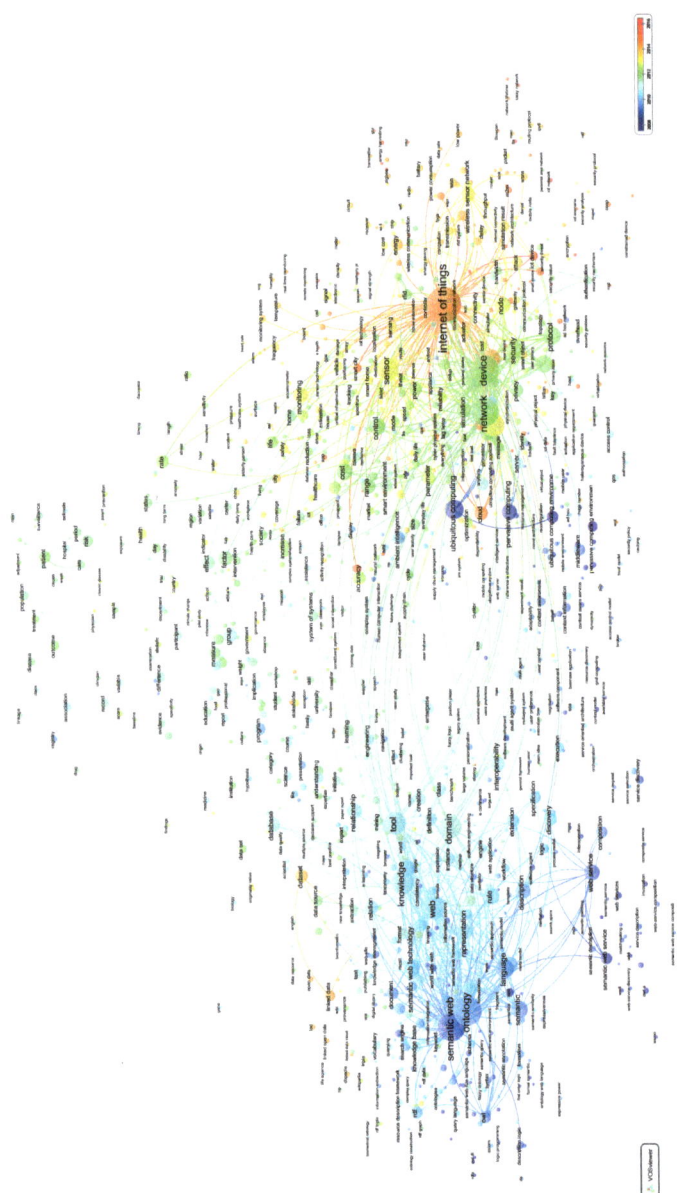

Fig. 2 A conceptual landscape, illustrating longitudinal development of the ten concepts 2008–2016 as well as other terms that are produced in connection with these. 1409 terms occurring at least 50 times in the material are shown. Large nodes indicate frequent usage of terms. Proximity between nodes illustrates the degree of connectedness across texts. Colder colors are reserved for usage in older publications (left side) while warmer colors indicate recent publications (right side)

This shift from theoretical to concrete concepts can indicate processes of research communities moving from vision to implementation over time. Such a development may not be surprising since these have been dramatic years involving breakthrough of the smartphone, app technology, broadband wireless Internet, social media, big data predictive analysis, and much more. The focus in the current section is on the many terms surrounding the 10 main concepts as they develop 2008–2016. Again, semantic web and Internet of things dominate the conceptual landscape of Fig. 2, leaving the other eight concepts with minor presence. This is partly due to the way that most of the concepts already had peaked by 2008. However, a more substantial explanation lies in the sheer numbers of publications associated with semantic web and Internet of things during 2008–2016.

6.3   The Five Dominating Domains

Looking closely at the various concepts highlighted in Fig. 2, it is interesting to note that of the conceptual domains identified in the previous section, five are clearly visible, while four exhibit low presence.

The first of the dominating domains is *smart*. On the left side of Fig. 2, many key concepts associated with the semantic web can be included in this domain: *knowledge, representation, understanding, semantic, language, syntax* and *discovery*. As we move to the right side of Fig. 2, the abstract concepts for intelligence are replaced by more concrete terms, i.e. *smart city, smart environment* and *smart home*. Crucially, more abstract and trendy concepts within the "academic smart domain" are missing such as *artificial intelligence, machine learning, deep learning* and *neural networks*.

The second of the dominating domains is *environment*. Curiously enough, this domain is largely missing on the left side of Fig. 2. Right of middle of the figure, we find the key concepts *smart environment* and *ubiquitous computing*. Here, the primary concern is the use of sensor technology which can be embedded in all forms of environment. On the right side of Fig. 2, we find concrete concepts such as *smart city, smart building, home* and *location*. In addition, there are also some distinct physical places such as *house, hospital* and *classroom*.

The third of the dominating domains is *tool*. This domain is largely missing on the left side of Fig. 2. On the right side, we find *device, receiver, chip, appliance,* and *wearable*. Notably, the tool domain is not used metaphorically, but rather signaling concrete things.

The fourth of the dominating domains is *networks*. On the left side of Fig. 2, the concept *web* has a strong presence not only through the semantic web but also through concepts such as *World Wide Web, web site, web service, web application* and *web technology.* There are also other network oriented terms present, such as *linked data, linked data cloud* and *linked open data.* On the right side of Fig. 2, the concept *network* has, seemingly, replaced *web* as the key concept in this domain. In addition, the concrete concept *wireless* becomes used: *wireless communication, and wireless sensor network.* The metaphorical concept *cloud* can also be seen to be situated in the network domain.

The fifth of the dominating domains is *control*. On the left side of Fig. 2, we find concepts such as *rule, access control* and *workflow.* Concepts within the control domain seem to be at the time concerned with managing code. At the right side we find concepts dealing with control of (smart) networks, devices and environments; *monitoring, controller, surveillance, safety, synchronization, authentication* and *security.*

The above five domains can be seen as illustrating various areas of technological expertise. Although we have placed various concepts within different domains, there are instances of overlap where some concepts could be situated within two domains.

6.4   The Four Domains Less Visible

Although they were recognized in the review of the other eight concepts, four domains appeared less visible in our inventory of associated concepts during the period 2008–2016. These are domains one would expect many stakeholders coming from a non-technological background to be concerned with. Given that the material investigated is primarily technological in character, their absence is certainly not surprising. However, given that we have been investigating concepts concerned with visualizing a technologically rich future within all of the Web of Science databases, including research from the humanities and the social sciences, the absence of these domains is, nonetheless, striking.

The first less visible domain is *user.* There are a few concepts connected to humans in Fig. 2. Those are uniformly represented as small nodes, indicating less frequent usage. Humans do figure in some professional roles or relations: *expert, student, participant, elderly person, stakeholder, physician,* and *patient.* The lack of strong concepts within the user domain signals a technology first perspective, that technology rich futures are built, so to speak, from hard to soft. From a CTA perspective it is troubling to find a wealth of concrete concepts in the last years of our material, but, with limited terms situated within the user domain.

The *society* domain is largely missing. Merely *government* and *society* are present with small nodes, signaling lack of societal perspectives. Policy concepts such as *regulation*, *legislation* and *compliance* are not found in the material.

The third less visible domain is *movement*. There is not much to be found in this domain on the left side of Fig. 2. However, as we focus on the right side of Fig. 2, we find a few concrete concepts: *GPS* and *vehicle*. The absence of the movement domain is puzzling as this material covers a time period of revolution for everyday mobility involving wireless, mobile and broadband Internet.

The *business* domain is largely missing. There are a few exceptions, *business application, enterprise and ecommerce* have small nodes but there is little to be found beyond that. In this material, it is not the economy, but the technology that drives visions of the future. As was demonstrated through our review of influential concepts, it was clear that most concepts investigated had been constructed by ICT researchers and this can explain the lack of business oriented vocabulary. A few concepts had roots in the private sector and work within multinational corporations such as IBM, Philips, AT&T and Xerox. Notably, these four different corporations are providers of "hard technology" and not representatives of the "new digital economy". Corporations such as Facebook, Google and Apple may prefer to push for conceptual innovations of their own making in the future.

7 Conclusion

The overarching problem that this article has dealt with is how the scrutiny of stakeholders utilizing the notions of CTA would be able to connect with relevant concepts for the discussion of technology rich futures. In our investigation of such concepts, we have identified and discussed three different issues concerning concepts: diversity, domains and lifecycle.

There is an obvious wealth and diversity of concepts describing technology rich futures. We have identified and described the 10 most frequently used of these, as found in the Web of Science databases. In so doing, we have also discussed different problems. The *fragmentation problem* is a consequence of different stakeholders attaching their reading and reflection to separate concepts. Therefore, critical scrutiny becomes clustered around various concepts and unconnected with each other. The *pioneering problem* follows up on the fragmentation problem as those attempting to find previous reflections regarding technology rich futures tend to be stuck with the literature surrounding one concept.

Finding a striking lack of previous reflections, such actors take on the role of pioneers, i.e., in a sense, reinventing the wheel.

We also identified nine conceptual domains, discussing how different concepts could allow for separate ways of talking about technology. This could be seen as a kind of *domain fragmentation*. When reviewing conceptual domains in the material specifically collected for 2008–2016, it became obvious that terms associated with "softer domains" (user, society, movement and business) exhibited low visibility. This is particularly notable as we identified a clear shift of concepts in recent years becoming more concrete, latched on to specific innovative projects.

Reviewing the lifecycle of concepts (Figs. 1 and 2) we found that these were similar to hype cycles (Fenn and LeHong 2011), quickly expanding in usage and within a few years settling down to a relatively modest number of yearly publications. Obviously, such shifts in conceptual trends create problems for stakeholders. We discussed the difficulty of nonspecialized stakeholders often becoming connected to discussions of the concept at a relatively late stage, when ICT researchers already may be on the verge of migrating to another concept.

The year 2008 stands out as marking a shift in our material. From this time on, the semantic web and the Internet of things appear to have taken on dominating positions as key concepts for discussions on technology rich futures. That said, it was also interesting to analyze the development of numerous specialized and more concrete concepts in recent years. This highlights a risk that stakeholder discussions will be focused on everyday technological issues such as security, privacy and authentication rather than the broader concepts articulating visions of technology rich futures.

We also noted the lesser visibility of conceptual domains associated with interests of non-technological stakeholder perspectives: *user, society, movement* and *business*. All of these were visible in some of the eight concepts taking on a lower profile after 2008 and the increasing dominance of Internet of things. No doubt, the Internet of things has become a powerful concept which can be used to talk about controlling a wide variety of concepts. It has become a way to deal with the *control of smart tools with the help of smart networks in smart environments*. In addition, it is obvious from our material that it is useful as a base for promoting more concrete concepts involving specific projects. In this way, the Internet of things take on the role of an umbrella concept, fruitful in generating concrete language that can be promoted as being part of a larger project.

However, there is always a choice to be made in the development of strong concepts. In the development of visions of technological rich futures it matters what kind of concepts are used to talk about the future. The heavy emphasis on Internet and smart networked relationships to things, rather than to humans, signal a need for other umbrella concepts of use to talk about technology rich futures. The choice to emphasize technological concepts, again and again, implies a choice to deselect conceptual starting points of humans; their relationships, practices, organizations and governments.

References

Ashton, K. (2009). That 'Internet of things' thing-In the real world, things matter more than ideas. *RFID Journal* (22 June). http://www.rfidjournal.com/articles/view?4986. Accessed 2 Jan 2018.

Augusto, J. C., & Aghajan, H. (2009). Editorial: Inaugural issue. *Journal of Ambient Intelligence and Smart Environments, 1*(1), 1–4.

Badger, M., Bushmitch, D. Agnish, V., Cozby, R. Fikus, J., Halloran, F., & Chang, K. et al. (2012). Laboratory-based end-to-end network system of systems integration, design and risk reduction: Critical activity for system of systems integration directorate and the army. military communications conference, 1–6. https://doi.org/10.1109/milcom.2012.6415712.

Berners-Lee, T., Hendler, J., & Lassila, O. (2001). The semantic web. *Scientific American* (May), 29–37.

Billsus, D., Brunk, C. A., Evans, C., Gladish, B., & Pazzani, M. (2002). Adaptive interfaces for ubiquitous web access. *Communications of the ACM, 45*(5), 34–38.

Bizer, C., Heath, T., Idehen, K., & Berners-Lee, T. (2008). Linked data-The story so far. In T. Heath, M. Hepp, & C. Bizer, C. (Eds.), *Special issue on linked data, International Journal on Semantic Web and Information Systems (IJSWIS)*. http://tomheath.com/papers/bizer-heath-berners-lee-ijswis-linked-data.pdf. Accessed 2 Jan 2018.

Brumitt, B., Meyers, B., Krumm, J., Kern, A., & Shafer, S. (2000). Easyliving: Technologies for intelligent environments. In H. W. Gellersen (Ed.), *Handheld and ubiquitous computing* (pp. 97–119). Berlin: Springer.

Cook, D. J., & Das, S. K. (2005). *Smart environments-Technologies, protocols, and applications*. Hoboken: Wiley.

Douma, K. F., Karsenberg, K., Hummel, M. J., Bueno-de-Mesquita, J. M., & van Harten, W. H. (2007). Methodology of constructive technology assessment in health care. *International Journal of Technology Assessment in Health Care, 23*(2), 162–168.

Ducatel, K., Bogdanowicz, M., Scapolo, F., Leijten, J., & Burgelman, J.-C. (2001). *Scenarios for ambient intelligence in 2010-Final report*. Seville: Institute for Prospective Technological Studies. https://cordis.europa.eu/pub/ist/docs/istagscenarios2010.pdf. Accessed 2 Jan 2018.

Fenn, J., & LeHong, H. (2011). Hype cycle for emerging technologies. Gartner Research 2011. https://www.gartner.com/doc/1754719/hype-cycle-emerging-technologies-.

Finkelstein, A. C. W., Savigni, A., Kappel, G., Retschitzegger, W., Kimmerstorfer, E.,Schwinger, W., & Feichtner, C. (2002). *Ubiquitous web application development-A framework for understanding*. Paper presented at the 6th World Multiconference on Systematics, Cybernetics and Informatics (SCI), Orlando, FL. http://www.cs.ucl.ac.uk/staff/A.Finkelstein/papers/uwa.pdf. Accessed 2 Jan 2018.

Garfield, E. (1979). *Citation indexing: Its theory and application in science, technology, and humanities.* New York: Wiley.

Genus, A. (2006). Rethinking constructive technology assessment as democratic, reflective, discourse. *Technological Forecasting and Social Change, 73,* 13–26.

Gluhak, A., Bauer, M., Montagut, F., Stirbu, V., Johansson, M., Vercher, J. B., et al. (2009). Towards an architecture for a real world internet. In G. Tselentis, J. Domingue, A. Galis, A. Gavras, D. Hausheer, S. Krco, V. Lotz, & T. Zahariadis (Eds.), *Towards the future internet: A European research* (pp. 313–324). Amsterdam: IOS Press.

Guinard, D., & Trifa, V. (2009). *Towards the web of things: Web Mashups for embedded devices.* Paper presented at the Workshop on Mashups, Enterprise Mashups and Lightweight Composition on the Web (MEM 2009). Madrid, Spain. http://webofthings. org/2009/04/20/web-mashups-mem/. Accessed 2 Jan 2018.

Guston, D. H., & Sarewitz, D. (2002). Real-time technology assessment. *Technology in Society, 24*(1), 93–109.

Huang, C. M., Jang, M. Y., & Chao, Y. C. (2000). CTW: An integrated computer and telephone-accessed WWW system. *Software: Practice and Experience, 30*(13), 1485–1507.

Kelley, M., & Pei, R. (1999). C4I2WS system of systems integration for the future army. In N. L. Faust, & S. Kessinger (Eds.), *Modeling, simulation, and visualization for real and virtual environments.* Papers Presented at Aerosense, 3694 (pp. 215–222). Orlando: SPIE Press.

Kessler, M. M. (1963). Bibliographic coupling between scientific papers. *American Documentation, 14*(1), 10–25.

Knichel, D. G. (2010). Life after future combat system, a family of ground robotic systems. In G. R. Gerhart, D. W. Gage, & C. M. Shoemaker (Eds.), *Unmanned Systems Technology XII, 7692.* FL: SPIE.

Krueger, M. W. (1977). Responsive environments. In *Proceedings of the June 13–16, 1977, National computer conference* (pp. 423–433). ACM, New York.

Lakoff, G., & Johnson, M. (2008). *Metaphors we live by.* Chicago: University of Chicago press.

Liu, C., Zhou, X., & Orlowska, M. E. (1998). *Issues in work flow and web-based work flow systems.* Paper presented at the 1998 Asia Pacific Web Conference (APWeb98), Beijing, 27–30 September. http://www.it.swin.edu.au/personal/cliu/APWeb98.ps. Accessed 2 Jan 2018.

Logenthiran, A., Pratiwadi, R., Logenthiran, D., Porebski, A., & Thomas, D.W. (1998). *Software migration of telecommunication network management systems to the web using CORBA and Java.* Paper presented at the thirty-first Hawaii International Conference on System Sciences, Kohala Coast, HI, 6–9 January. http://ieeexplore.ieee.org/ stamp/stamp.jsp?tp=&arnumber=649265. Accessed 2 Jan 2018.

Marquet, L. C., & Ratches, J. A. (1998). Future directions of information systems in the army after next. In R. Suresh (Ed.), *Digitization of the Battlespace III, 3393* (pp. 20–26). Orlanda: SPIE Press.

Moens, N. P., Broerse, J. E., Gast, L., & Bunders, J. F. (2010). A constructive technology assessment approach to ICT planning in developing countries: Evaluating the first phase, the roundtable workshop. *Information Technology for Development, 16*(1), 34–61.

O'Reilly, T. (2005). What is Web 2.0? Design patterns and business models for the next generation of software. http://www.oreilly.com/pub/a/web2/archive/what-is-web-20. html. Accessed 2 Jan 2018.

Olson, N., Nolin, J. M., & Nelhans, G. (2015). Semantic web, ubiquitous computing, or internet of things? A macro-analysis of scholarly publications. *Journal of Documentation, 71*(5), 884–916.

Orwat, C., Graefe, A., & Faulwasser, T. (2008). Towards pervasive computing in health care-A literature review. *BMC Medical Informatics and Decision Making, 8*(1), 26.

Rip, A. (2008). Nanoscience and nanotechnologies: Bridging gaps through constructive technology assessment. In G. Hirsch Hadorn, H. Hoffmann-Riem, S. Biber-Klemm, W. Grossenbacher-Mansuy, D. Joye, C. Pohl, U. Wiesmann, & E. Zemp (Eds.), *Handbook of transdisciplinary research* (pp. 145–157). Berlin: Springer.

Saha, D., & Mukherjee, A. (2003). Pervasive computing: A paradigm for the 21st century. *Computer, 36*(3), 25–31.

Satyanarayanan, M. (2001). Pervasive computing: Vision and challenges. *IEEE Personal Communications, 8*(4), 10–17.

Schot, J. (1992). Constructive technology assessment and technology dynamics: The case of clean technologies. *Science, Technology and Human Values, 17*(1), 36–56.

Schot, J., & Rip, A. (1996). The past and future of constructive technology assessment. *Technological Forecasting and Social Change, 54,* 251–268.

Small, H. (1973). Cocitation in scientific literature-New measure of relationship between 2 documents. *Journal of the American Society for Information Science, 24*(4), 265–269.

Sundmaeker, H., Guillemin, P., Friess, P., & Woelfflé, S. (Eds.). (2010). *Vision and challenges for realising the internet of things.* Brussels: European Commission, Information Society and Media.

van Eck, N. J., & Waltman, L. (2010). Software survey: VOSviewer, a computer program for bibliometric mapping. *Scientometrics, 84*(2), 523–538.

van Raan, A. F. (2017). Sleeping beauties cited in patents: Is there also a dormitory of inventions? *Scientometrics, 110*(3), 1123–1156.

Weiser, M. (1991). The Computer for the 21st century-Specialized elements of hardware and software, connected by wires, radio waves and infrared, will be so ubiquitous that no one will notice their presence. *Scientific American* (September), 94–104.

Weiser, M. (1993a). Some computer science issues in ubiquitous computing. *Communications of the ACM, 36*(7), 75–84.

Weiser, M. (1993b). Ubiquitous computing. *Computer, 26*(10), 71–72.

Weiser, M. (1995). The last link: The technologist's responsibilities and social change. *Computer-Mediated Communication Magazine, 2*(4), 17.

Weiser, M., Gold, R., & Brown, J. S. (1999). The origins of ubiquitous computing research at PARC in the late 1980s. *IBM Systems Journal, 38*(4), 693–696.

Yachir, A., Amirat, Y., Chibani, A., & Badache, N. (2012). Towards an event-aware approach for ubiquitous computing based on automatic service composition and selection. *Annals of Telecommunications-Annales des Télécommunications, 67*(7/8), 341–353.

York, J., & Pendharkar, P. C. (2004). Human-computer interaction issues for mobile computing in a variable work context. *International Journal of Human-Computer Studies, 60*(5–6), 771–797.

Zelkha, E., & Epstein, B. (1998). *From devices to 'Ambient intelligence': The transformation of consumer electronics.* Paper presented at the digital living room conference, Philips, June 12.

Jan Nolin (Ph.D.) is professor and head of the research group Social Media Studies as well as head of research at the Swedish School of Library and Information Science, Borås, Sweden.

Gustaf Nelhans (Ph.D.) is senior lecturer at the SSLIS at University of Borås, Sweden, focusing on the performativity of scientometric indicators as well as on the theory, methodology and research policy aspects of the scholarly publication in scientific practice, specifically on the evaluation of societal relevance such as professional impact.

Nasrine Olson (Ph.D.) is a senior lecturer/researcher, at the Swedish School of Library and Information Science (SSLIS), University of Borås, Sweden. She is a co-leader of the research group Social Media Studies and the coordinator of the EU funded project SUIT-CEYES.

Intervening Through Futures for Sustainable Presents: Scenarios, Sustainability, and Responsible Research and Innovation

Lauren Withycombe Keeler, Michael J. Bernstein and Cynthia Selin

Abstract

Discourses around innovation often unreflexively assume positive progress and the inevitable contribution of new technologies to the betterment of society. Little attention is paid to issues of sustainability—including intergenerational equity, justice, and socio-ecological integrity—and the complex ways that societal arrangements and sociotechnical regimes are intermingled. Innovation governance *for sustainability* needs to actively engage both responsible research and innovation and sustainability paradigms in order for science and technology to effectively serve societal and sustainability goals. There is an opportunity to utilize tools of foresight to raise the capacity of actors in innovation processes to consider alternative framings of progress and challenge the status quo. This chapter explores participatory scenario construction as a

L. W. Keeler (✉) · M. J. Bernstein
School for the Future of Innovation in Society, Arizona State University, Tempe, USA
e-mail: Lauren.Withycombe@asu.edu

M. J. Bernstein
e-mail: mjbernst@asu.edu

C. Selin
School for the Future of Innovation in Society & School of Sustainability,
Arizona State University, Tempe, USA
e-mail: Cynthia.Selin@asu.edu

© Springer Fachmedien Wiesbaden GmbH, part of Springer Nature 2019   255
A. Lösch et al. (eds.), *Socio-Technical Futures Shaping the Present*,
Technikzukünfte, Wissenschaft und Gesellschaft / Futures of Technology,
Science and Society, https://doi.org/10.1007/978-3-658-27155-8_12

means to productively disrupt status-quo imaginaries. The Future of Wastewater Sensing, a participatory scenario study, is presented as a case example to inform sustainability-oriented responsible research and innovation.

1   Introduction

Sustainability science stems from a normative position that current, dominant trajectories of innovation are undesirable and untenable for the long-term viability of people and the planet (WCED 1987). The field focuses on urgent problems broaching the balancing of economic, environmental, and societal goals, where knowledge and action are necessary in the short- to medium-term to avert crossing critical thresholds or planetary boundaries vital to the survival and thriving of life on Earth (Röckstrom et al. 2009). There is explicit concern about a gap between the perceived urgency with which deep sustainability transitions may be needed, and the actual pace of social and cultural (as opposed to technological) change (van der Leeuw et al. 2012). Values and norms often take generations to change, which may simply be too long to wait for addressing urgent sustainability challenges like climate change (Inglehart 2008). Since its inception sustainability science has emphasized the central role of science and technology to achieving sustainability goals (Clark and Dickson 2003). Despite acknowledgement of the limits of technological solutionism by some (e.g. Kemp 1994; Beder 1996; Hueseman 2003), scholars and practitioners continue to focus attention on the ways in which technological solutions can be deployed, nudging behaviour and forcing change despite obdurate cultural or political institutions. Efforts in sustainability thus can often be seen as calling for more rapid, explicitly technological solutions to energy, water, pollution, transportation, and hosts of other sustainability challenges (c.f., the 17 UN Sustainable Development Goals, each of which contain a science and technology component [United Nations 2015]). Technological solutions to sustainability challenges rest implicitly on a faith in the power of human ingenuity and entrepreneurship to innovate our way out of the unsustainable present and into a better future.

Unfortunately, technological solutions, while perhaps offering quicker fixes, also bring forth other dilemmas. Grossly speaking, technological innovations often spur unintended consequences, neglect how burdens and rewards are unequally distributed, or how they may accelerate ecological destruction (Jasanoff 2016). Attempts to innovate out of problems (Nelson 2004) often lead to the creation of new ones (Westley et al. 2011). Mounting evidence suggests that technological outcomes of innovation often contribute to or exacerbate many of the challenges and inequities plaguing societies (c.f., Cozzens et al.

2005; Woodhouse and Sarewitz 2007; Owen et al. 2012; Forsberg et al. 2015). Dominant social paradigms for innovation, such as triple-helix arrangements of industry, government, and research organizations (Leydesdorff and Etzkowitz 1998), perpetuate neglect of these shortfalls and exclude broader social groups and organizations (Foley and Wiek 2013; Wiek et al. 2016b). A recent example of the broken promises of current technological configurations can be seen in the limitations of nanotechnology in addressing development and sustainability challenges (Cozzens et al. 2013; Wiek et al. 2012), despite such claims being central to justifications for substantial investments in research and development (Salamanca-Buentello et al. 2005). A longer-lived example can be found in the unintended consequences of fossil fuel dependence on environmental, health, and social outcomes (Tainter and Taylor 2014).

A key dilemma for sustainability has thus become how to, through mediating technological change, influence the slow changing, "press" dynamics (Collins et al. 2011), of social and cultural change. There is precedent for this enterprise, given the underlying ways in which technologies are shaped by and shape human morals and thus values (Sweirstra and Rip 2007). Indeed, the challenge is how to raise forth the human and the cultural when encountering decisions about technological innovation, so often shrouded in and dominated by a calculus of speed, risk, and machinery. Scenario research methods in particular offer a means of surfacing moralities and values often underlying technological change (Selin 2008; Boenink et al. 2010) yet crucially do so in a present-focused modality. That is, scenarios are crafted in the present to reimagine technological futures so that, in the present, different choices can be made. In this chapter, we will explore participatory scenario construction as one means of surfacing the values and norms key for sustainability that serve to enrich the modes of attention and range of present choices about configurations of technological change. In revealing a case study associated with technology development, we demonstrate how creating spaces of intersection with sustainability imaginaries help support responsible research and innovation.

1.1   Innovation Governance Challenges

To govern innovation is at its core an effort to grapple with promoting long-term integrity of people and the planet without succumbing to unintended, undesirable consequences (Merton 1936). Innovation entails activities and actions of diverse social groups to create (or alter) material artifacts with intention (Pinch and Bijker 1987). Some have argued that innovation cannot be directed and can only be regulated after the fact (Polanyi 1962). A growing body of scholars, policy makers, and practitioners recognize that innovation is shaped through the individual and collective expression of values held and decisions made by people involved in innovation activities (c.f. Williams and Edge 1996; Hekkert et al. 2007). Innovation governance is made difficult by three overarching quandaries: (i) *why* pursue innovation (*orientation*); (ii) *who* to involve in this process and why (*legitimacy*); (iii) *how* to manage innovation to achieve a desired outcome (*control*).

Underlying each of these innovation governance quandaries are the set of imaginaries that people hold about the present and future. Status-quo orientation of innovation in western industrialized countries are replete with dreams of leisure, wealth, and happiness realized through material acquisition and greater efficiency. In this mode, innovation depends on privilege and exploits implicit and explicit bias by leveraging legacies of power built on oppression. These forces combined with fear of reprisal interpenetrate the power grasped by those who claim legitimacy in efforts to lead, spur, manage, or otherwise direct innovation. Resulting inventions inevitably serve to reinforce existing social and economic structures. From logics of legitimacy through privilege follow exploitative and dehumanizing efforts to control innovation processes in pursuit of materially-mediated enlightenment/progress. Confronting such status-quo beliefs, thoughts, and actions is a paramount challenge for sustainability. Indeed, the capability of human kind to shape innovation for a sustainable present—and future—is a battleground for how technologies are shaped, what values are imputed into material artefacts through social processes, and what social orders infuse or are imposed on the world as a result.

1.2   Fostering Responsible Research and Innovation. A Capacity-Building Approach

Following this logic, sustainability values and norms, if brought into conversation with status-quo socio-technical imaginaries sustained through powerful regimes and cultural norms, might provide one avenue for supporting responsible research and innovation alternatives.

Direct attempts by scholars of science, technology, and innovation to intervene in the practices of research and innovation can be seen as an effort in capacity building according to Hankins (2013). Capacity building is summoned to draw attention to the civic and interpersonal skills and knowledge brought to bear in examining and choosing among different innovation pathways. It is a recognition that the social, cognitive, and emotional skills required to question and cope with turbulent, uncertain, systemic, and novel situations need honing. Such an effort can have multiple audiences, including publics (Selin et al. 2017) and expert communities (Bernstein et al. 2017; Withycombe Keeler et al. 2017). Continuing with the example of publics, Selin et al. (2017) focused on the temporal dimensions of the capacity of citizens to confront socio-technical change. They considered which sorts of interventions create space for citizens to act with greater "regard to potential futures, based on contemporary observations, buttressed by past experiences" (p. 10). This example notwithstanding, such an explicit orientation to the future, with intentions to open up alternatives and better account for complex change, is often lacking in efforts to advance responsible research and innovation.

Indeed, the same is true in efforts to advance sustainability. Integrated assessment efforts for sustainability (related to emerging science and technology domains) have also been found to lack holistic visions, narratives, and identification of values related to the future (Forsberg et al. 2016). While future-oriented assessments may have an anticipatory stance, they often stem from reductionist, non-systemic perspective on the future, and lack inclusion of robust anticipatory methods (Forsberg et al. 2016). Further, they are dominated by approaches focused on risk mitigation whereas sustainability demands a complementary, constructive approach: one capable of identifying opportunities for improvement of innovation. In the remainder of this chapter we explore the potential of participatory scenario construction to integrate key capacities from responsible research and innovation and sustainability as a means to productively grapple with innovation governance quandaries.

2 Linking Responsible Research and Innovation Capacity to Sustainability Competencies

For responsible research and innovation (RRI) to contribute to sustainability and benefit from the insights produced through sustainability science, the complementarity and differences between necessary capacities need explication. From technology assessment and its formative and constructive antecedents come anticipation, reflexivity, and engagement (sometimes referred to as inclusion) as a framework for enabling actors within the innovation process to implement responsible research and innovation (Barben et al. 2008; Stilgoe et al. 2013; von Schomberg 2013; Owen et al. 2012; Guston 2014). *Anticipation* broadly explores plausible futures and their implications for innovation. This includes considering how science and technology might be changed by society but also how innovation changes society by driving shifts in morals, norms, and behaviors. *Reflexivity* requires reflection and metacognition on the actions that take place during the innovation process. As Fisher and colleagues (2006) have exploited with "midstream modulation", reflexivity is the creating of space for and practice around questioning why particular decisions are taken. For the governance of science, technology, and innovation, reflexivity is critical to uncovering how individual motivations and cultural norms manifest in the minutia of scientific and technical practice (which can ultimately impact the trajectory and consequences of innovations). Finally, engagement refers to bringing diverse publics into the innovation process and allowing the upstream assessment of science and technology to include those who pay for it (through their taxes or investments) and those whose lives will ultimately be impacted.

This general framework for responsible research and innovation processes overlaps significantly with key competencies for sustainability research and problem-solving, yet with some helpful additions. Wiek and colleagues (2011a, b) specify five competencies that when integrated enable individuals and teams to understand and address sustainability problems. They include, (i) systems thinking, (ii) values thinking, (iii) futures thinking, (iv) strategic thinking, and (v) collaborative competence. Systems thinking is the ability to identify variables and interactions that make up complex social-ecological-technical systems in order to understand how they function. This includes thinking and integrating across spatial and governance scales (e.g. local, global), societal domains (e.g. politics, economics), and system types (e.g. social, ecological). Values thinking is the ability to assess the sustainability of present and future systems and identify, analyze, and assess tradeoffs in the implementation of sustainability efforts. It also includes the identification, integration, and negotiation of different and conflicting values around sustainability problems and their potential solutions. Futures thinking is the ability to craft alternative futures that explore how current social-ecological-technical systems (or aspects thereof) could evolve in the future and to create, in collaborative settings, sustainability visions that are systemic, reflect sustainability principles, integrate diverse values, and can provide target knowledge to guide strategic planning and policy making. Strategic thinking is the ability to design and implement sustainability transitions including the crafting of evidence-based plans, policies and programs that can realize sustainability goals over time and account for barriers and the carriers necessary to overcome them. Finally, collaborative competence is the set of interpersonal skills necessary to work with diverse groups to identify sustainability problems and their root causes, envision sustainable futures and explore alternatives, and create strategies to achieve sustainable futures that have broad buy-in. These competencies undergird more recent work on transformational capacity, or the ability of individuals, organizations and systems to catalyze transformational change (Wolfram et al. 2016; Wiek et al. 2019).

Against this backdrop, we offer the process and products of participatory scenario construction as a means of productively integrating these competencies and capacities, acknowledging the truly transformational potential of responsibly governed technological innovation for sustainability.

3   Scenarios: Futures and Present

Both sustainability and responsible research and innovation emphasize shared ownership of futures in the making. Scientists share with politicians, regulators, funders, and other innovators in the division of responsibility and moral labor for their research and innovations (Rip 2014; Stilgoe et al. 2013). Similarly, scientists, politicians, activists, and communities (among others) share responsibility for ensuring the integrity of socio-ecological systems for future generations. Investigation and elaboration of how such responsibility manifests and leads to more societally beneficial and sustainable research outcomes remains a core task for science, technology, and innovation scholars (Lindner et al. 2016a).

Within responsible innovation and anticipatory governance, scenario methodology has been used and proposed as one means of fostering the development of anticipatory and reflexive capacities across enactors of innovation (Guston 2008; Barben et al. 2008; Rip and Kulve 2008; Selin 2011; Owen et al. 2013; Stilgoe et al. 2013; Guston 2014). Scenario methods have also been embraced by sustainability science because of the way the method can be used to explore complex phenomena ridden with uncertainty, incorporate perspectives and expertise from different groups, integrate qualitative and quantitative data, and inform present action on sustainability issues (Swart et al. 2004). While there are a variety of scenario construction methods, from quantitative extrapolation to exploratory (van Oost et al. 2016) to those focused on probability or plausibility (Ramírez and Selin 2014), a common theme of each is the importance of knowledge integration. Scenarios provide a sequence of steps and, in the case of participatory scenarios, the time and space to bring to bear different and otherwise detached perspectives.

Scenarios can also expand notions of what is governed and by whom in the innovation process—moving from a concept of governance as risk mitigation and regulation, to one in which uncertainties, motivations, purpose, social and political context and plausible innovation trajectories are made transparent and open to reflection, adaptation and capacity building as necessary (Selin 2007; Stilgoe et al. 2013). Beck (2000) points to organized irresponsibility as a challenge to realizing responsible research and innovation and an expanded concept of innovation governance. The complexity of the innovation process and the number of actors and competing interests involved make it difficult to identify responsible parties and hold them accountable. Scenarios as a method of foresight and knowledge integration can make some of these complexities transparent, making it possible to identify discreet actors and entities who could take responsibility or stand in for aspects of the innovation process. Such an approach differs from, for example, hard-law regulatory regimes of liability, reflecting instead a means of fostering, in the case of RRI, tangible and actionable responsiveness to "ethically acceptable and socially desirable interests" (von Schomberg 2013; Arnaldi et al. 2016). Scenario construction can enable potentially responsible parties to consider plausible future outcomes and respond in kind, through capacity building, but also, if upstream in the innovation process, through changes to research processes and/or technologies under development (Barben et al. 2008; Stilgoe et al. 2013).

The focus is not exclusively on anticipation, however, which misses the practical value of scenario development and the need to integrate and build an array of capacities to make an expanded concept of innovation governance possible. Structured and participatory scenario construction does explore the future, but in such a fashion so as to reframe the present, inviting reflexive and strategic responses (Ramirez and Wilkinson 2016). This reframing proves valuable for clarifying ill-structured, wicked problems and contested areas where the "answers" are more ambiguous but action is necessary nonetheless. By exploring alternative futures, better questions in the present can be formulated. Scenario construction is about redrawing the frames of reference under scrutiny when encountering and questioning innovation.

This re-framing brings more variables into consideration than those normally captured by status-quo discourses on innovation. By painting portraits of future worlds, scenario users are invited to systemically consider values, societal regimes, governance patterns and culture. "A fundamental presupposition of STS is that any imagination of new technologies is necessarily coupled with the shaping or even inventing of novel kinds of users and usages, of altered social relations or even new societal regimes" (see the editors' introduction to this volume). In this way, by enlarging the frames of attention, participatory scenario creation provides space for exploring alternatives and opening up assumptions often hidden in popular discourse.

This refocusing of attention is a matter of perception, yet changed perception influences action. Once new realities are disclosed, once frames have been realigned, different scopes of action and choices are revealed. This re-orientation is critical for sustainability and historical patterns of action and the societal arrangements that enabled them are ill-suited to creating fundamentally different futures that meet sustainability goals. However, while a straightforward or generic accounting of the performativity of scenario development is regularly postulated, taking the present–oriented impacts out of context is risky and misleading. There are no reliable formulaic approaches to scenario development once one takes into account the context of use, disparate purposes, themes under inquiry, eclectic actors, and different degrees of finesse in implementation. There is no future visioning factory where inputs and outputs are standardized. The achievements of scenario development—like the future itself—are necessarily contingent on being in the room where it happens. Recognizing this importance in the particulars, we offer a case on Wastewater Sensing to illustrate the ways that scenarios can revitalize attention on sustainability in all its richness and complexity.

4   Case Study: The Future of Wastewater Sensing

In its last year, the Center for Nanotechnology in Society, a National Science Foundation-funded center charged with exploring and attempting to grapple with some of the innovation governance quandaries around nanotechnology, had the opportunity to conduct a series of scenario workshops with science and engineering partners at Arizona State University (ASU). The workshops were part of a pivot from 10 years of research focused on the anticipatory governance of nanotechnology to a future agenda that leveraged those insights to refine and innovate ways of making responsible research and innovation a reality across a spectrum of research areas.

One such workshop focused on the Future of Wastewater Sensing (WWS) and was conducted in partnership with ASU's Center for Environmental Security and the NSF-funded LC Nano. These centers have ongoing research into WWS technology and its potential to reveal public health information and recover resources from wastewater streams. In particular, the workshop focused on the public health aspect of wastewater sensing and the ability of mass-spectrometry to sense in wastewater a variety of contaminants harmful to humans and the environment (Venkatesen and Halden 2014; Venkatesen et al. 2015). Venkatesen and Halden (2014) (both of whom participated in the workshop) hypothesized that "[wastewater treatment plants] can serve as chemical observatories to study the prevalence and likely fate of chemicals and their bioaccumulation potential in human society and the environment." The scenario workshop, The Future of Wastewater Sensing Technology, explored critical uncertainties surrounding the development and dissemination of wastewater sensing (WWS) technology and its potential to be of societal benefit and contribute to environmental sustainability.

WWS is part of a larger movement in public health to leverage large data sets ("big data") to better understand the underlying causes of observed, population-level health problems and predict the onset and spread of illness and disease (Khoury and Ioannidis 2014). Such epidemiological applications largely exist as socio-technical imaginaries as significant technological, social, and legal barriers remain to making sense of and responding effectively to such enormous amounts of data (Lazer et al. 2014). For WWS specifically, the Center for Environmental Security has created the United States Municipal Sewage Sludge Repository that as of 2014 contained 164 samples from different municipalities across the US. The repository is open access and the center's leadership has invited all interested and capable to query the data (Venkatesen et al. 2015). There is plenty to find in the sludge. Beyond environmental contaminants, sewage sludge can contain illicit drugs, alcohol, antibiotics, patent-protected chemotherapy drugs and anything else consumed and discharged or runoff into the sewer-shed of a community (Lazer et al. 2014). Such broad application invites a number of social, legal, and ethical questions that need to be considered as part of a responsible research and innovation paradigm.

As a suite of technologies and applications at different stages of the innovation process, WWS has the potential to fall through the analytical cracks that exist in any single sustainability or responsible research and innovation intervention approach. One could analyze the contribution of WWS to narrow sustainability-related objectives, such as identifying environmental contaminants, and analyze the value of the technology relative to existing wastewater treatment and environmental sensing approaches. Alternatively, one could invite reflection on why particular contaminants are being sensed, by whom, and to what end to try to understand what specific governance challenges are posed by this research. Each analysis could build important capacities among those involved in the WWS innovation system but each is necessarily incomplete. The Future of Wastewater Sensing Workshop provided on opportunity to integrate responsible research and innovation and sustainability paradigms and hold space for considering alterative futures with significant implications for the present.

The Future of Wastewater Sensing Workshop was held November 2–3, 2015 at Arizona State University in Tempe, Arizona, USA. In attendance were members of the Center for Environmental Security and LC Nano, including PIs, municipal water providers, social scientists, STS scholars, military personnel (interested in combat and humanitarian applications), local law enforcement, a federal regulator (EPA), and legal scholars.

The workshop utilized the scenario axes or intuitive logics technique developed by the Royal Dutch Shell Company in the late 60 s and early 70 s, later refined by the Global Business Network in the 90 s and in the last decade, the Oxford School of Scenario Planning (Ramirez and Wilkinson 2016). To illustrate the potential of scenarios for responsible research and innovation, generic steps of the intuitive logics method are presented, followed by the specified application with wastewater sensing, the capacities for sustainability and responsible research and innovation that are built, and the contribution to a broader concept of innovation governance (summarized in Table 1).

4.1   Selecting Clients, Defining Focal Question and Selecting Participants

Intuitive logics begins by defining a focal group or client for whom the scenarios are created. In this way, the method is explicitly action-oriented, creating anticipatory knowledge (Wiek et al. 2011a) that is intended to be integrated into decision-making in the present. In this case the 'client' was the Center for Environmental Security (CES) and LC Nano specifically and wastewater sensing research at ASU broadly. Dr. Rolf Halden, the director of CES and our partner in the scenario workshop was explicitly interested in attending to the social implications of the technology and informing research and development so as to contribute to environmental sustainability and public health without unleashing negative, unintended consequences. In defining the focal question, interviews were conducted with Dr. Halden and Dr. Paul Westerhoff, the PI of LC Nano researching the potential for resource recovery from wastewater sludge. Interviews were also conducted with all workshop participants to help refine the focus of the workshop. Workshop participants were identified by the 'clients' and the CNS research team. The process of defining a timeframe and focal question and selecting participants already begins to introduce reflexivity and systems thinking. All interviewees were asked about key factors to consider with WWS and critical uncertainties affecting the future of the technology. They were also asked about analogous technologies that might hold lessons or cautionary tales of WWS. The interview protocol was directed at unearthing assumptions about how such an envisioned technology interacts with society, who it impacts and benefits, and what lessons might be garnered from analogous technologies and doing so from multiple perspectives.

Table 1 How scenario construction builds and aligns responsible research and innovation capacities and sustainability competencies to enable innovation governance as exemplified through the wastewater sensing workshop

Step in scenario construction	Case study: The future of WWS	Contribution to innovation governance (Stilgoe et al. 2013)	RRI capacity built (Guston 2014)	Sustainability competence built (Wiek et al. 2011a)
Define focal questions and timeframe	What might wastewater sensing technology and its applications look like in 2035 based on uncertainties surrounding future changes in economic, social, political, and technological domains?	Articulating the "purposes" of innovation" and "motivations for innovation"	Reflexivity	Systems thinking
Identify participants	WWS Researchers Legal Scholars STS and Ethics Scholars Regulators (EPA) Water Managers Military	Articulating the "purposes" of innovation" Understanding the social and political context of innovation	Engagement	Systems thinking; Collaborative competence
Brainstorm driving forces	Selected examples: Reality of water scarcity Public perception of sensing Liability	Grappling with uncertainty	Anticipation	Systems thinking; Futures thinking

(continued)

Table 1 (continued)

Step in scenario construction	Case study: The future of WWS	Contribution to innovation governance (Stilgoe et al. 2013)	RRI capacity built (Guston 2014)	Sustainability competence built (Wiek et al. 2011a)
Identify critical uncertainties	Selected examples: What is being sensed? Who owns the data? Who is doing the sensing and for what purpose?	Grappling with uncertainty	Anticipation and Reflexivity	Futures thinking; Values thinking
Select scenario axes	Ownership (Open-source/public v. Private) What is sensed (Individual v. Community)	Trajectories and directions of innovation	Anticipation and Reflexivity	Futures thinking; Values thinking
Sketch scenario storylines	Four Scenarios: End 2 End Sensing Bio.Me Technocrats for the Community Precious Bodily Fluids	Trajectories and directions of innovation	Anticipation	Systems thinking; Futures thinking
Write scenario narratives	*See Fig. 1 for scenario features	Trajectories and directions of innovation	Anticipation	Systems thinking; Futures thinking
Assess scenarios	Winners and Losers SWOT Analysis	Motivations of Innovation	Reflexivity	Values thinking; Collaborative competence

(continued)

Table 1 (continued)

Step in scenario construction	Case study: The future of WWS	Contribution to innovation governance (Stilgoe et al. 2013)	RRI capacity built (Guston 2014)	Sustainability competence built (Wiek et al. 2011a)
Proposals for Action	Talk to CDC about WWS Do cost-benefit analysis vis a vis WWS and public health threats Create guide to communicate about concentrations of toxins found through WWS and what people can do	Social and political constitutions of innovation	Reflexivity	Strategic thinking

4.2 Brainstorming Driving Forces

Identifying the client, focal questions and timeframe, and participants and conducting interviews are all essential pre-work to the participatory scenario construction workshop. In the Future of Wastewater Sensing workshop, the first major step of the scenario construction process was to identify driving forces that will impact the future of WWS technology *in society*. Participants were encouraged to consider broader contextual factors and trends outside the locus of control of any individual in the room. The process of brainstorming driving forces is highly generative; facilitators push participants to think across a broad array of domains (e.g. social, technological, political, environmental) and to specify how they envision the trends impacting WWS. The identification and specification of these driving forces is an exercise in systems thinking and futures thinking or anticipation. Workshop participants grappled with the systems that are implicated by and impact the potential technology. Some of these relationships are evident in the present, others are speculations about relationships that will emerge or change in the future.

4.3 Identifying Critical Uncertainties and Selecting Scenario Axes

Once the participants identify all driving forces they can muster or were worth noting, these driving forces are sorted to identify critical uncertainties by situating them relative to one another by their uncertainty and their potential impact on the focal question. This was a valuable opportunity to deliberately inflect scenario construction with ideas from RRI and sustainability. In the case of the WWS workshop participants were encouraged to identify critical uncertainties that mattered for sustainability and for ensuring that the technology is societally beneficial. Two examples of critical uncertainties included: (1) *Who is doing the sensing?* Participants felt that a municipal government sensing wastewater to ensure public health would have different outcomes for the technology and society than an insurance company sensing wastewater to determine premiums for a particular zip code; and (2) *Who owns the data?* Participants were concerned about the affordability and data burden of the technology leading to the outsourcing of analysis of wastewater data and the potential for communities to be targeted by any number of groups as a result of private companies owning sensed data.

From the critical uncertainties, axes were selected that served as the scaffolding to create scenarios in a two-by-two matrix. Selected axes in the wastewater sensing workshop were: (1) Who owns the data (Public/Open source v. Private)? and (2) What is being sensed (Individuals v. Communities)? For the latter axis, participants felt that societal response and governance challenges would be fundamentally different depending on the scale of the sensing. Participants also felt that the ability of the technology to achieve its intended effects was highly dependent on what was being sensed. The culling of driving forces, identification of critical uncertainties, and final selection of axes are analytical and normative exercises. From both sustainability and responsible research and innovation perspectives, this process can build capacity for values thinking and promote reflexivity by having diverse participants communicate about what is important to consider vis a vis the technology or entity under consideration. It can also build capacity for futures thinking and anticipation as participants make arguments for why particular driving forces are more uncertain than others and speculate as to the relative plausibility of a range of alternative outcomes.

4.4 Sketching Scenarios and Writing Scenario Narratives

Sketching scenario storylines and writing scenario narratives captures and communicates the complexity of alternative futures. From the scenario axes participants speculated on the content of each of the four scenarios, considering what wastewater sensing might look like in 2035 (See Fig. 1 for Summary).

The sketching step was done in plenary to ensure a shared understanding of the scenarios and that once breakout groups formed, the scenarios they each took would evolve to be sufficiently distinct. During the sketching, participants were prompted to consider the impact of the scenarios for sustainability and responsible research and innovation considerations. In particular, participants were asked, what are the implications of each scenario for power structures, for environmental and public health, and for vulnerable populations, and what kind of governance regimes might emerge in each scenario. Scenarios can be constructed without this kind of framing but for integrating responsible research and innovation and sustainability this is an opportunity to provide a normative orientation to the process without inserting the researchers own expectations into the content of the scenarios. After the scenarios were sketched, breakout groups formed to tackle individual scenarios.

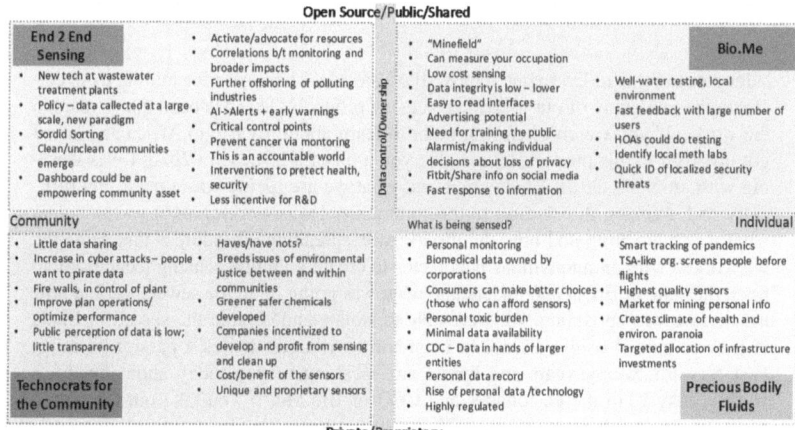

Fig. 1 Four scenarios of wastewater sensing

4.5   Scenario Narratives and Relevant Key Features

Narratives can be a means of articulating systems understandings. Participants were pushed to consider some of the systemic changes (social, economic, legal, etc.) necessary to make their scenarios plausible. In writing narratives participants engaged in a kind of unstructured, participatory consistency analysis by negotiating their different understandings of how particular aspects of their scenarios fit together and reinforced one another. In this step participants actively engage in anticipation, considering how a plausible future technology could shape and be shaped by society. Once the narratives were completed participants read or performed them in plenary. Through the narrative writing, participants not only engaged with anticipation and built futures thinking and systems thinking competence, they also built interpersonal competence by grappling with how to best communicate their understanding of this future and laid the foundation for further engagement of other stakeholders or publics. Below are excerpts from the scenario narratives developed by participants and some of those scenario key features and early warnings that participants felt were important to consider and reflect upon.

4.5.1 Bio.Me Scenario

"Good evening, I'm Dr. [name removed] of Bio.Me. Thank you for joining us to celebrate the launching of our individual sensing products. This has been many years in the works. If you weren't here last night to learn about my trip to Africa and how I got involved in this process, let me give you a bit of a history. In 2015, I was working with an NGO on the Ebola crisis and what we are seeing was a persistent difficulty with tracking the disease. In the community I was working in, it was not clear what was happening and how the disease was spreading. To address this data gap, we worked with an anonymous benefactor to create a mobile sensing technology to track the state of Ebola and understand what was going on in the sewershed and how the disease was spreading. We were able to isolate and contain the spread of Ebola to prevent spread of the disease to neighboring villages. It was a passion of mine. Fast forward twenty years and here I am speaking to you today about my life's work. Today I am the co-founder and CEO of Bio.Me. If you take out the period here, we are looking at a biome with our technology. It's about entire living communities. Not only plants, or animals, but also human communities. My company is generously backed by private donors and we have also partnered with the military to deploy sensing technology in war and humanitarian crises...."

Key Features

- Commercialization of wastewater sensing technology.
- Public ownership of data does not preclude private sensing for corporate, military, or law-enforcement uses.
- Broad public interest in personal sensing but quality of sensing linked to affordability.
- Wastewater sensing closely associated with individual health monitoring.
- Movement from community scale sensing to individual sensing.
- Individual access to their data, broad public data literacy.

Early Warnings (that we might be headed toward this future)

- Continued interest in individual health monitoring devices (FitBit).
- Public acceptance of community scale wastewater sensing.
- Increased public comfort with wastewater (Reduced yuck factor).
- The quantified self grows in popularity.
- Educational programs target youth, training them in monitoring their own health.

4.5.2 End2End Sensing for the Community

"This is the lead-up to the Centers for Disease Control (CDC) report presentation from the Principal Investigator on the preliminary results from the pilot National Wastewater Sensing Center, a 10-city network. The network leader is putting together a major report demonstrating the progress of the network to the CDC. She is fielding calls from city presenters, talking though challenges with wastewater treatment plant interfaces and resource recovery progress; her talking points will include carbon sequestration credits, fertilizer offsets, quality product, and GOLD. ... There is an urgency to get this pilot off the ground. ...The scene opens in a hallway and there you see a woman with an earpiece. It's the network leader wearing her iEar. She is talking to her New York City coordinator who just got off of a very angry call with the Texas legislature who told him to shove his waste somewhere else because they don't have any more land to give to New York for their sludge..."

Key Features of the Scenario

- The scale of human waste from cities makes sensing and recovery essential but there is still public resistance and resistance from elected officials.
- Some cities are implementing wastewater sensing.
- Wastewater sensing is seen by some as being important to sustainability.
- Sensing focuses on public and environmental health, it is about grappling with problems of urbanization, population growth and environmental contamination.
- More catastrophic storms, disease outbreaks, bring sustainability issues front and center.
- Inter-state fights about waste and waste exporting.
- Safety issues with waste recovery and waste transfer are front and center.

Early Warnings (that we may be headed toward this future)

- Push to keep organics out of landfills, a tightening of environmental regulations around waste and wastewater.
- Increased interest in waste as a commodity.
- Waste seen as a national problem. Pressure from the environmental groups.
- Federal interest in sensing for public health.
- Bottom up adoption of wastewater sensing by municipalities.

4.5.3 Technocrats for the Community

"...One day the town of Greensburg starts detecting something new in the water. This [wastewater sensing] technology allows them to see illicit drugs and identify when new drugs are introduced to the community. The facility found some sort of new particle that was created due to the combination with a nanoparticle and reflaxin, known as nanoflaxin. There has been an increase in dead fish in the neighboring village, Bluesville, where the wastewater flows. There is a predominance of this new particle in the river as well. The researchers at the university found that there is a correlation between this particle and the deformities in the fish and the increased hospitalization of Bluesville children. ... University of Greensburg, the wastewater treatment facility, and the hospital begin coordination. They have a large database of information they have been collecting from the wastewater treatment facility...There is no information getting out to the people because the data is privately owned and proprietary. A whistleblower at the university comes forward and... tells the public that researchers have made a connection between something being discharged from the Greensburg wastewater treatment plant and illnesses that are arising downstream in Bluesville."

Key Features

- Plausible, likely continuation of current trends.
- Fully developed wastewater sensing technology integrated into municipal systems of wealthier communities.
- Data ownership and privacy make it difficult to alert neighboring communities about issues of public health.
- Demands for transparency from the community.
- Wastewater "haves" and "have nots."

Early Warnings (that we are headed for this future)

- Adoption of wastewater sensing technology by municipalities.
- Limited transparency of data collected from wastewater sensing.
- Public interest in detecting more environmental toxins in wastewater.
- Wastewater sensing proposed as a solution to issues like municipal water contamination events.
- Interest in wastewater sensing from regulatory agencies like EPA or CDC.

4.5.4 Precious Bodily Fluids

Alice cursed while she frantically fussed in the kitchen. "They'll be here any minute! Bob, be a dear and make sure the bathroom is locked."

"I'm decanting the Pinot, Al. I want it to breathe. The last Alaskan red we had needed more time."

"Just don't forget."

Examining her face and hair in one of the many reflective surfaces in the kitchen, she asked, "Bob, how do I look?"

"Simply radiant dear. Following the advice of your rEliLilly Poo really agrees with you." Bob went on to arrange the cheese plate redolent with gorgonzola. He was startled by the sudden chiming of the doorbell.

There is a couple hosting a dinner party and before the guests arrive, they are making sure to lock the bathroom door. They are part of the movement called "Not in My Toilet," a futuristic version of Not in My Backyard where they don't want someone's foreign waste to come in and influence the data in their waste stream. The folks who are coming to visit them are part of a minority group and the national waste census is the next day and they want to make sure they get counted. So, there is a conflict over counting waste.

Key Features of this Future

- Commoditization of waste(water). Company "20 pee and me."
- Large aggregation of data.
- National and international conflicts over sensing practices and what is detected. Exacerbated by mass migration due to climate change.
- New movement of "Not in my toilet," which is similar to current movement "Not in my back yard." This creates conflicts.
- Public Effluence and Accommodation Act gives the government authority to collect data from wastewater streams as part of the national census.
- Because waste has value and we have things like the waste census, people want to and feel obliged to contribute.
- People cannot be constrained around providing their contribution so if you are a restaurant you have to allow people to come off the street and use your bathroom so that their contribution can be counted.

Early Warnings

- Normalization of "potty talk" – people discuss bodily functions as part of normal conversation.
- Large migration and demands from countries to better monitor health of displaced peoples.
- Investment by private companies in wastewater sensing technology.
- Interest in mining of wastewater sludge and resource recovery.

4.6 Assessing Scenarios and Developing Proposals

Once the scenarios have been crafted and presented there is an opportunity to once again explicitly address issues of concern for responsible research and innovation and sustainability. In assessing the scenarios, it is possible to look at each of their implications for sustainability and speculate as to ways in which decisions could be taken in the present by actors in the innovation process to promote societally beneficial and sustainable outcomes with and from a technology. In the Wastewater Sensing Workshop a modified SWOT (Strengths, Weaknesses, Opportunities, and Threats) analysis was done in plenary. Participants were asked about the strengths and weaknesses of WWS technology across the scenarios, where they saw opportunities for researchers to advance WWS in a responsible way, and what kind of threats they saw that might impede the responsible and sustainable development of the technology. The workshop concluded by inviting participants to make proposals for how the clients or anyone else in the room could help advance wastewater sensing in a responsible manner. These last steps are critical as they link futures thinking and anticipation with strategic thinking and concrete actions linked to specific actors so that an expanded concept of innovation governance can be made actionable.

5 Discussion and Outlook

This chapter explored the potential for scenario construction to advance a combination of responsible research and innovation and sustainability capacities for improved technology governance. This is proposed as part of broader effort by scholars of science, technology, and innovation, as well as other interested communities, to intervene in the practices of research and innovation in an effort to build capacity for sustainability. Innovation systems can be obdurate, and therefore building capacity to implement responsible research and innovation, oriented and informed by sustainability, cannot happen through a single intervention. Rather, scenario construction was offered as one important tool in the toolbox of those interested in aligning science and technology with societal and sustainability goals. Addressing the innovation governance quandaries proposed in this paper requires incentive structures in innovation systems so that they may better realize normative outcomes articulated in sustainability and responsible research and innovation discourses, among them justice, socio-ecological integrity, and human flourishing (Foley et al. 2016). Scenarios are a useful tool for bringing the socio-technical imaginaries created in expert, science-policy spaces into conversation with very different conceptualizations of desirable and plausible futures. However, sustained engagement is necessary to overcome regression to status-quo and effectively integrate different perspectives and actively engage in anticipation and reflection throughout innovation processes.

Sustainability values and norms, when brought into conversation with status-quo socio-technical imaginaries, provide one avenue for supporting responsible research and innovation around alternatives. However, this requires a strong, shared concept of sustainability to drive the conversation toward issues of long-term societal concern. In the case of the WWS workshop, participants struggled to reconsider their own roles in the process and tended to try to find ways to make wastewater sensing work. The scenario construction process was not designed to manufacture support for WWS technology but rather to critically reflect on how the technology could develop in unexpected ways with an eye toward steering research and development toward sustainable ends. The difficulty of navigating systemic complexities to chart sustainable development of a technology may reflect, in part, the fact that a common, operational definition of sustainability was intentionally omitted from the workshop. This decision was out of pragmatic concern for logistics (not enough time) and priorities (the clients were interested in the scenario construction process, not an informed conversation about dimensions of sustainability).

Engaging in participatory scenario construction as a part of the innovation process brings to the foreground the social nature of innovation and the embeddedness of science and technology in society. Making space for deliberation about the future is essential for experts and publics alike. The scenario construction method discussed in this chapter works well for smaller groups with higher levels of knowledge about the focal question. Futures methods need to be further tested to better understand what they are and are not capable of delivering. For example, it is unclear whether such participatory scenario construction workshops centered on a technology can effectively make space for the possible consideration that a technology ought not be developed at all.

Through deliberation on socio-technical futures shaping the present, scenarios enable the redrawing of what's in and what's out in an analysis of innovation pathways. Through scenarios, stories are created which invite consideration on the values and morals and cultures that shape innovation. Further, as researchers of science, technology, and innovation continue to grapple with innovation governance quandaries, scenario construction can help groups propose and vet possible answers to questions of *why* to pursue innovation, *whom* to involve, and *how* to manage for desirable and sustainable outcomes.

Acknowledgement This research was undertaken with support by the Center for Nanotechnology in Society of Arizona State University (CNS-ASU),funded by the U.S. National Science Foundation (cooperative agreement #0531194 and #0937591). The findings and observationscontained in this article are those of the authors and do not necessarily reflect the views of the U.S. National Science Foundation.

References

Arnaldi, S., Gorgoni, G., & Pariotti, E. (2016). RRI as a governance paradigm: What is new. In R. Lindner, S. Kuhlmann, S. Randles, B. Bedsted, G. Gorgoni, E. Giessler, A. Loconto, & N. Mejlgaard (Eds.), *Navigating towards shared responsibility in research and innovation approach* (pp. 23–29). Karlsruhe: Fraunhofer ISI.

Barben, D., Fisher, E., Selin, C., & Guston, D. H. (2008). Anticipatory governance of nanotechnology: Foresight, engagement, and integration. In J. Hackett & O. Amsterdamska (Eds.), *The handbook of science and technology studies* (3rd ed., pp. 979–1000). Cambridge: MIT Press.

Beck, U. (2000). The cosmopolitan perspective: Sociology of the second age of modernity*. *The British Journal of Sociology, 51*(1), 79–105. https://doi.org/10.1111/j.1468-4446.2000.00079.x.

Beder, S. (1996). *The nature of sustainable development*. Newham: Scribe Publications.

Bernstein, M. J., Reifschneider, K., Bennett, I., & Wetmore, J. M. (2017). Science outside the lab: Helping graduate students in science and engineering understand the complexities of science policy. *Science and Engineering Ethics, 23*(3), 861–882.

Boenink, M., Swierstra, T., & Stemerding, D. (2010). Anticipating the interaction between technology and morality: A scenario study of experimenting with humans in bionanotechnology. *Studies in Ethics, Law, and Technology, 4*(2), 1–28.

Clark, W. C., & Dickson, N. M. (2003). Sustainability science: The emerging research program. *Proceedings of the National Academy of Sciences, 100*(14), 8059–8061.

Collins, S. L., Carpenter, S. R., Swinton, S. M., Orenstein, D. E., Childers, D. L., Gragson, T. L., et al. (2011). An integrated conceptual framework for long-term social–ecological research. *Frontiers in Ecology and the Environment, 9*(6), 351–357. https://doi.org/10.1890/100068.

Cozzens, S. E., Bobb, K., Deas, K., Gatchair, S., George, A., & Ordonez, G. (2005). Distributional effects of science and technology-based economic development strategies at state level in the United States. *Science and Public Policy, 32*(1), 29–38.

Cozzens, S., Cortes, R., Soumonni, O., & Woodson, T. (2013). Nanotechnology and the millennium development goals: water, energy, and agri-food. *Journal of Nanoparticle Research, 15*(11), 1–14.

Fisher, E., Mahajan, R. L., & Mitcham, C. (2006). Midstream modulation of technology: Governance from within. *Bulletin of Science, Technology & Society, 26*(6), 485–496.

Foley, R. W., & Wiek, A. (2013). Patterns of nanotechnology innovation and governance within a metropolitan area. *Technology in Society, 35*(4), 233–247.

Foley, R. W., Bernstein, M. J., & Wiek, A. (2016). Towards an alignment of activities, aspirations and stakeholders for responsible innovation. *Journal of Responsible Innovation, 3*(3), 209–232.

Forsberg, E.-M., Quaglio, G., O'Kane, H., Karapiperis, T., Van Woensel, L., & Arnaldi, S. (2015). Assessment of science and technologies: Advising for and with responsibility. *Technology in Society, 42,* 21–27.

Forsberg, E.-M., Ribeiro, B., Heyen, N. B., Nielsen, R., Thorstensen, E., de Bakker, E., et al. (2016). Integrated assessment of emerging science and technologies as creating learning processes among assessment communities. *Life Sciences, Society and Policy, 12*(1), 9.

Guston, D. H. (2008). Innovation policy: Not just a jumbo shrimp. *Nature, 454*(7207), 940–941.

Guston, D. H. (2014). Understanding 'anticipatory governance'. *Social Studies of Science, 44*(2), 218–242.

Hankins, J. (2013). Endnotes: Building capacity for responsible innovation. In R. Owens, J. Bessant, & M. Heinz (Eds.), *Responsible innovation: Managing the responsible emergence of science and innovation in society* (pp. 269–273). London: Wiley.

Hekkert, M. P., Suurs, R. A. A., Negro, S. O., Kuhlmann, S., & Smits, R. E. H. M. (2007). Functions of innovation systems: A new approach for analysing technological change. *Technological Forecasting and Social Change, 74*(4), 413–432.

Huesemann, M. H. (2003). The limits of technological solutions to sustainable development. *Clean Technologies and Environmental Policy, 5*(1), 21–34.

Inglehart, R. F. (2008). Changing values among western publics from 1970 to 2006. *West European Politics, 31*(1–2), 130–146.

Jasanoff, S. (2016). The floating ampersand: STS past and STS to come. *Engaging Science, Technology, and Society, 2,* 227. https://doi.org/10.17351/ests2016.78.

Kemp, R. (1994). Technology and the transition to environmental sustainability: The problem of technological regime shifts. *Futures, 26*(10), 1023–1046.

Khoury, M. J., & Ioannidis, J. (2014). Big data meets public health. *Science, 346*(6213), 1054–1055.

Lazer, D., Kennedy, R., King, G., & Vespignani, A. (2014). The parable of Google Flu: Traps in big data analysis. *Science, 343*(6176), 1203–1205.

Leydesdorff, L., & Etzkowitz, H. (1998). Triple Helix of innovation. *Science and Public Policy, 25*(3), 195–203.

Lindner, R., Daimer, S., Beckert, B., Heyen, N., Koehler, J., Tuefel, B., et al. (2016a). *Addressing directionality: Orientation failure and the systems of innovation heuristic. Towards reflexive governance.* Karlsruhe: Fraunhofer ISI.

Lindner, R., Kuhlmann, S., Randles, S., Bedsted, B., Gorgoni, G., Griessler, E., et al. (2016b). *Navigating towards shared responsibility in research and innovation: Approach, process and results of the res-agora project.* Karlsruhe: Fraunhofer ISI.

Merton, R. K. (1936). The unanticipated consequences of purposive social action. *American Sociological Review, 1*(6), 894–904.

Nelson, R. R. (2004). The market economy, and the scientific commons. *Research Policy, 33*(3), 455–471.

Owen, R., Macnaghten, P., & Stilgoe, J. (2012). Responsible research and innovation: From science in society to science for society, with society. *Science and Public Policy, 39*(6), 751–760.

Owen, R., Stilgoe, J., Macnaghten, P., Gorman, M., Fisher, E., & Guston, D. (2013). A framework for responsible innovation. In R. Owen, J. Bessant, & M. Heintz (Eds.), *Responsible innovation: Managing the responsible emergence of science and innovation in society* (pp. 27–50). London: Wiley.

Pinch, T., & Bijker, W. E. (1987). The social construction of facts and artifacts: Or how the sociology of science and the sociology of technology might benefit each other. In W. E. Bijker, T. P. Hughes, & T. Pinch (Eds.), *The social construction of technological systems: New directions in the sociology and history of technology* (pp. 17–50). Cambridge: MIT Press.

Polanyi, M. (1962). The republic of science: Its political and economic theory. *Minerva, 1*(1), 54–73.

Ramírez, R., & Selin, C. (2014). Plausibility and probability in scenario planning. *Foresight, 16*(1), 54–74. https://doi.org/10.1108/FS-08-2012-0061.

Ramírez, R., & Wilkinson, A. (2016). *Strategic reframing: The Oxford scenario planning approach*. Oxford: Oxford University Press.

Rip, A. (2014). The past and future of RRI. *Life sciences, society and policy, 10*(1), 1–17. https://doi.org/10.1186/s40504-014-0017-4.

Rip, A., & Kulve, H. T. (2008). Constructive technology assessment and socio-technical scenarios. In C. Selin, E. Fisher, E. Wetmore, & M. Jameson (Eds.), *The yearbook of nanotechnology in society: Vol. I. Presenting futures*. Berlin: Springer.

Rockström, J., Steffen, W., Noone, K., Persson, A., Chapin, F. S., Lambin, L., et al. (2009). Planetary boundaries: Exploring the safe operating space for humanity. *Ecology and Society, 14*(2), 32.

Salamanca-Buentello, F., Persad, D. L., Court, E. B., Martin, D. K., Daar, A. S., & Singer, P. A. (2005). Nanotechnology and the developing world. *PLoS Med, 2*(5), e97. https://doi.org/10.1371/journal.pmed.0020097.

Selin, C. (2007). Expectations and the emergence of nanotechnology. *Science, Technology and Human Values, 32*(2), 196–220.

Selin, C. (2008). The Sociology of the Future: Tracing Stories of Technology and Time. *Sociology Compass 2* (6), 1878–1895. https://doi.org/10.1111/j.1751-9020.2008.00147.x.

Selin, C. (2011). Negotiating plausibility: Intervening in the future of nanotechnology. *Science and Engineering Ethics, 17*(4), 723–737.

Selin, C., Rawlings, K. C., de Ridder-Vignone, K., Sadowski, J., Altamirano Allende, C., Gano, G., Davies, S. R., & Guston, D. H. (2017). Experiments in engagement: Designing public engagement with science and technology for capacity building. *Public Understanding of Science*, 26 (6), 634–649. https://doi.org/10.1177/0963662515620970.

Stilgoe, J., Owen, R., & Macnaghten, P. (2013). Developing a framework for responsible innovation. *Research Policy, 42*(9), 1568–1580. https://doi.org/10.1016/j.respol.2013.05.008.

Swart, R. J., Raskin, P., & Robinson, J. (2004). The problem of the future: Sustainability science and scenario analysis. *Global Environmental Change, 14*(2), 137–146. https://doi.org/10.1016/j.gloenvcha.2003.10.002.

Swierstra, T., & Rip, A. (2007). Nano-ethics as nest-ethics: Patterns of moral argumentation about new and emerging science and technology. *NanoEthics, 1*(1), 3–20.

Tainter, J. A., & Taylor, T. G. (2014). Complexity, problem-solving, sustainability and resilience. *Building Research and Information, 42*(2), 168–181. https://doi.org/10.1080/096 13218.2014.850599.

United Nations. (2015). Addis Ababa Action Agenda of the third international conference on financing for development. http://www.un.org/esa/ffd/ffd3/wp-content/uploads/ sites/2/2015/07/Addis-Ababa-Action-Agenda-Draft-Outcome-Document-7-July-2015. pdf. Accessed 4 Juli 2017.

Van der Leeuw, S., Wiek, A., Harlow, J., & Buizer, J. (2012). How much time do we have? Urgency and rhetoric in sustainability science. *Sustainability Science, 7*(1), 115–120.

van Oost, E., Kuhlmann, S., Ordóñez-Matamoros, G., & Stegmaier, P. (2016). Futures of science with and for society: Towards transformative policy orientations. *Foresight, 18*(3), 276–296.

Venkatesan, A. K., & Halden, R. U. (2014). Wastewater treatment plants as chemical observatories to forecast ecological and human health risks of manmade chemicals. *Scientific Reports, 4,* 3731.

Venkatesan, A. K., Done, H. Y., & Halden, R. U. (2015). United States national sewage sludge repository at Arizona State University – A new resource and research tool for environmental scientists, engineers, and epidemiologists. *Environmental Science and Pollution Research, 22*(3), 1577.1586. https://doi.org/10.1007/s11356-014-2961-1.

von Schomberg, R. (2013). A vision of responsible research and innovation. In R. Owen, J. R. Bessant, & M. Heintz (Eds.), *Responsible innovation: Managing the responsible emergence of science and innovation in society* (pp. 51–74). London: Wiley.

WCED, World Commission on Environment and Development. (1987). *Our common future, from one earth to one world.* Oxford: Oxford University Press.

Westley, F., Olsson, P., Folke, C., Homer-Dixon, T., Vredenburg, H., Loorbach, D., et al. (2011). Tipping toward sustainability: Emerging pathways of transformation. *AMBIO, 40*(7), 762–780. https://doi.org/10.1007/s13280-011-0186-9.

Wiek, A., Withycombe, L., & Redman, C. L. (2011a). Key competencies in sustainability: A reference framework for academic program development. *Sustainability Science, 6*(2), 203–218.

Wiek, A., Withycombe, L., Redman, C., & Mills, S. B. (2011b). Moving forward on competence in sustainability research and problem solving. *Environment, 53*(2), 3–13.

Wiek, A., Foley, R. W., & Guston, D. H. (2012). Nanotechnology for sustainability: What does nanotechnology offer to address complex sustainability problems? *Journal of Nanoparticle Research, 14*(9), 1–20.

Wiek, A., Bernstein, M., Foley, R., Cohen, M., Forrest, N., Kuzdas, C., et al. (2016a). Operationalising competencies in higher education for sustainable development. In M. Barth, G. Michelsen, M. Rieckmann, & I. Thomas (Eds.), *Routledge handbook of higher education for sustainable development* (pp. 241–260). London: Routledge.

Wiek, A., Foley, R. W., Guston, D. H., & Bernstein, M. J. (2016b). Broken promises and breaking ground for responsible innovation–intervention research to transform business-as-usual in nanotechnology innovation. *Technology Analysis & Strategic Management, 28*(6), 639–650.

Wiek, A., Withycombe, Keeler L., Beaudoin, F., et al. (2019). Building transformational capacity for implementing sustainability solutions in urban areas. *Ambio, 48*(5), 494–506.

Williams, R., & Edge, D. (1996). The social shaping of technology. *Research Policy, 25*(6), 865–899.

Withycombe Keeler, L., Gabriele, A., Wiek, A., & Kay, B. (2017). Future shocks and city resilience: Building organizational capacity for resilience and sustainability through game play and ways of thinking. *Sustainability: The Journal of Record, 10*(5), 282–292.

Wolfram, M., Frantzeskaki, N., & Maschmeyer, S. (2016). Cities, systems and sustainability: Status and perspectices for research on urban transformations. *Current Opinion in Environmental Sustainability, 22*, 18–25.

Woodhouse, E., & Sarewitz, D. (2007). Science policies for reducing societal inequities. *Science and Public Policy, 34*(2), 139–150.

Lauren Withycombe Keeler (Ph.D.) is Assistant Professor and foresight practicioner in the School for the Future of Innovation in Society at Arizona State University where she is also a member of the Center for the Study of Futures and a Fellow in the Risk Innovation Lab.

Michael J. Bernstein (Ph.D.) is a an Assistant Research Professor based in Tempe, Arizona, at the School for the Future of Innovation in Society. He conducts transdisciplinary research in science and technology systems, supporting transition experiments for sustainability in domains spanning STEM higher education, national research policy, and corporate practices.

Cynthia Selin (Ph.D.) is an interdisciplinary scholar of Science, Technology and Society focused on foresight and the governance of innovation. She is an Associate Professor at the School for the Future of Innovation in Society and the School for Sustainability at Arizona State University where she directs the Center for the Study of Futures.

Part IV
White Paper on Technology Assessment and Socio-Technical Futures

Technology Assessment of Socio-Technical Futures—A Discussion Paper

Andreas Lösch, Knud Böhle, Christopher Coenen,
Paulina Dobroc, Arianna Ferrari, Reinhard Heil, Dirk Hommrich,
Martin Sand, Christoph Schneider, Stefan C. Aykut,
Sascha Dickel, Daniela Fuchs, Bruno Gransche,
Armin Grunwald, Alexandra Hausstein, Karen Kastenhofer,
Kornelia Konrad, Alfred Nordmann, Petra Schaper-Rinkel,
Dirk Scheer, Ingo Schulz-Schaeffer, Helge Torgersen
and Alexander Wentland

This chapter contains the translation of the German version of a discussion paper
collectively developed in the context of a workshop of scholars from the fields of
technology assessment, innovation analysis, science and technology studies (STS)
and foresight. The German version has been published in December 2016 online (see:
http://www.itz.kit.edu/112.php) and was discussed widely in the German speaking TA
community.

A. Lösch, K. Böhle, C. Coenen, P. Dobroc, A. Ferrari, R. Heil, D. Hommrich, M. Sand,
and C. Schneider have been the editorial staff of the ITAS-project "Visions as socio-
epistemic practices" (https://www.itas.kit.edu/english/projects_loes14_luv.php).

A. Lösch (✉) · K. Böhle · C. Coenen · P. Dobroc · R. Heil · A. Grunwald · D. Scheer ·
C. Schneider
Institute for Technology Assessment and Systems Analysis, Karlsruhe Institute of
Technology, Karlsruhe, Germany
e-mail: andreas.loesch@kit.edu

K. Böhle
e-mail: knud.boehle@kit.edu

C. Coenen
e-mail: christopher.coenen@kit.edu

P. Dobroc
e-mail: paulina.dobroc@kit.edu

© Springer Fachmedien Wiesbaden Gmbh, part of Springer Nature 2019   285
A. Lösch et al. (eds.), *Socio-Technical Futures Shaping the Present*,
Technikzukünfte, Wissenschaft und Gesellschaft / Futures of Technology,
Science and Society, https://doi.org/10.1007/978-3-658-27155-8_13

R. Heil
e-mail: reinhard.heil@kit.edu

A. Grunwald
e-mail: armin.grunwald@kit.edu

D. Scheer
e-mail: dirk.scheer@kit.edu

C. Schneider
e-mail: christoph.schneider@mailbox.org

A. Ferrari
Department Strategy and Content, Futurium gGmbH, Berlin, Germany
e-mail: ferrari@futurium.de

D. Hommrich
Headoffice, German Council for Scientific Information Infrastructures,
Göttingen, Germany
e-mail: dirk.hommrich@rfii.de

M. Sand
Department of Values, Technology and Innovation, TU Delft,
BX Delft, The Netherlands
e-mail: m.sand@tudelft.nl

S. C. Aykut
Department of Socioeconomics, University of Hamburg, Hamburg, Germany
e-mail: stefan.aykut@wiso.uni-hamburg.de

S. Dickel
Institute for Sociology, Johannes Gutenberg University, Mainz, Germany
e-mail: dickel@uni-mainz.de

D. Fuchs · K. Kastenhofer · H. Torgersen
Institute of Technology Assessment, Austrian Academy of Sciences, Vienna, Austria
e-mail: daniela.fuchs@oeaw.ac.at

K. Kastenhofer
e-mail: kkast@oeaw.ac.at

H. Torgersen
e-mail: torg@oeaw.ac.at

B. Gransche
Institute of Advanced Studies - FoKoS, University of Siegen, Siegen, Germany
e-mail: bruno.gransche@uni-siegen.de

A. Hausstein
Institute of Technology Futures, Karlsruhe Institute of Technology,
Karlsruhe, Germany
e-mail: alexandra.hausstein@kit.edu

Summary

Problem: Visions of technology, future scenarios, guiding visions (Leitbilder) represent imaginations of future states of affairs that play a functional role in processes of technological research, development and innovation—e.g. as a means to create attention, communication, coordination, or for the strategic exertion of influence. Since a couple of years there is a growing attention for such imaginations of futures in politics, the economy, research and the civil society. This trend concerns technology assessment (TA) as an observer of these processes and a consultant on the implications of technology and innovation. TA faces increasing demands to assess imaginations of futures that circulate in the present and to participate in shaping these through scenarios or foresights. More than ever, this raises the question, which propositions can be made based on these imaginations by TA and how this can be used in advisory practices. Imaginations of futures are relevant for TA not as predictions but in their significance and effectiveness in the present, which need to be understood and assessed.

Contents: This discussion paper outlines how present significance and effects of imagined futures in technological research and innovation processes can be conceived and analyzed. In this paper, all forms of imaginations of technology futures will be called "socio-technical futures" because

K. Konrad
Department of Science, Technology and Policy Studies, University of Twente,
Enschede, The Netherlands
e-mail: k.e.konrad@utwente.nl

A. Nordmann
Institute of Philosophy, Technical University of Darmstadt, Darmstadt, Germany
e-mail: nordmann@phil.tu-darmstadt.de

P. Schaper-Rinkel
Center Innovation Systems & Policy, Austrian Institute of Technology,
Vienna, Austria
e-mail: Petra.Schaper-Rinkel@ait.ac.at

I. Schulz-Schaeffer
Department of Sociology, TU Berlin, Berlin, Germany
e-mail: schulz-schaeffer@tu-berlin.de

A. Wentland
Munich Center for Technology in Society, Technical University of Munich,
Munich, Germany
e-mail: alexander.wentland@tum.de

within them technological developments and social changes are interwoven and inseparably interrelated. In this paper, we discuss (1) why TA should analyze socio-technical futures, (2) how such analyses can grasp the societal conditions (e.g. power structures) that are expressed in the imagined futures and how these become effective in processes of technology development, communication, decision making etc. We raise the question (3) which self-reflexive positioning or possible realignment of TA is needed as a response to its increased concern with assessing and even co-producing socio-technical futures. The latter is often demanded regarding the growing attention by politics and publics to imaginations of futures with wide temporal and spatial reach.

Addressee of this paper is the TA community in a broader sense. The *aim* is to sensitize colleagues for the topic and its challenges, to consolidate discussions and to provide theoretical and methodical suggestions for research in TA and related advisory practices with respect to socio-technical futures. This paper has been originally initiated during the workshop "The present of technological futures-theoretical and methodical challenges for Technology Assessment" (March 2016, Karlsruhe), in which all of the paper's authors participated. The contents of this discussion paper are preliminary results that shall initiate and guide further discussions.

1   Introduction

To fulfil its purpose, i.e., estimating and assessing both intended and unintended consequences of technological developments and innovation processes for society and environment, technology assessment (TA) inevitably has to deal with disputed and sometimes highly speculative imaginations of the future. For more than ten years, an increasing commitment of the TA community with technology-related ideas of the future (such as future imaginaries, technological visions, scenarios, guiding visions) can be observed. We grasp all kinds of these imaginations regarding the future by the provisional term "socio-technical futures". This term does not refer to the reification of futures. We therefore speak of socio-technical futures as these futures always relate technological developments to social change. With a different range, these socio-technical futures envision connections between technological and social changes. They express the desires, fears, interests and preferences of their producers and users (see Sect. 2). The reason why TA is dealing increasingly with such socio-technical futures has been and is, on the one hand, the expanding orientation of TA towards new and emerging sciences and technologies (NEST). On the other hand, there is a growing demand e.g., of politics for orientational knowledge when it comes to decision-making processes regarding temporally far-reaching transformations. Respective examples are the energy transformation, the digitalization of society, measures against climate change or in the context of political support programs such as responsible research and innovation (RRI).

In its dealing with socio-technical futures for its research and advice purposes, which knowledge can be generated by TA at all? The imaginations of the future cannot actually predict what the future reality will be like. However, both public debates and debates among experts (whether in the mass media, in research policy, in the context of participation procedures or in the context of research and development) cannot be imagined without them. The necessity for TA to analyze and assess those imaginations results from the fact that their impact on research-political strategies or on public opinions cannot be denied. Socio-technical futures assert their significance in the present. This is why TA must know the roles, functions and consequences of such socio-technical futures when it comes to the processes of creating, spreading and using them. Thus TA is demanded to analyze these futures as an expression of the current society. This means to consider different and sometimes disputed interests, desires, preferences and claims to power of groups, which are part of our present society and which design, distribute and communicate socio-technical futures and finally base their decision-making and actions on them. In this context, TA has the task of putting narratives of "purely technological innovation" into question, of criticizing too narrow views of certain options for the future and of pointing out alternatives. But what does this imply for TA in terms of theory, methods, empirical work and its advisory practice?

This paper offers suggestions on how TA can deal with the theoretical, methodical and empirical challenges when it comes to its analysis and assessment of the significance of technology-related ideas of the future as socio-technical futures, and as codes and ways of expressing current societal states of affair. These challenges are manifold and multi-levelled. Dealing with them requires specific analytical approaches to be able to understand (see e.g., the debate on "hermeneutical TA") what such socio-technical futures might mean for current constitution and dynamics of society in connection to their respective contexts and processes of production and use as well as what might be related consequences. This all requires the reflected integration of the theories, methods and procedures of TA, and also of Science and Technology Studies (STS), and other disciplines of the humanities and the social sciences (see Sect. 3).

Thus, what the authors believe to be necessary is an original and TA-specific way of analyzing and assessing the present state of such socio-technical futures and the further development of the required methods of analysis and assessment. Dependent on the context (e.g., research, politics, development, mass media), technological visions-currently e.g., of big data, synthetic biology, open source or nursing robotics-do not only refer to different things, they also affect different processes (e.g. research, experiments, negotiations, planning or controversies) in different ways. Accordingly, the various challenges and consequences of making use of the relevant procedures for the advisory practices of TA as an interventionist science (e.g., as parliamentary TA, participative TA, constructive TA and as foresight) must be reflected upon. As TA is not just an observer and analyst of socio-technical futures but contributes-intentionally or unintentionally-to shaping them and intervenes in processes by means of its advice, the authors believe TA has to appropriately reflect on its own role and effects, and to position itself accordingly (see Sect. 4). The paper intends to support such an increased theoretical, empirical and methodical reflection by TA on present socio-technical futures as well as an appropriate self-reflection and positioning of TA when it comes to research and advice.

The focus is on why and how TA, in a theoretically and methodically reflected way, deals with socio-technical futures and on the resulting, necessary self-reflection of the practice of TA. The paper is divided into three parts:

1. Reasons are given why TA should deal with socio-technical futures as an expression of current society and with their efficacy;
2. it is shown how analyses of socio-technical futures may proceed methodically;
3. TA is situated as an actor involved in the processes under analysis.

The *addressees of the paper* are interested scientists from the TA community in a broader sense. Apart from advisory TA, which is oriented at intervening by way of advising political institutions and the governance of processes (e.g., parliaments, science, research and innovation policies), we also reckon other kinds of TA among this community/addressees. This involves foresights, futures research as well as innovation- and STS in the fields of humanities and social sciences if their research and advice performance deals with the analysis and assessment of technology-related ideas of the future (in the sense of socio-technical futures).

2   Why Should TA Analyze Socio-Technical Futures?

2.1   What are Socio-Technical Futures?

By socio-technical futures we mean projections and ideas, which-explicitly or implicitly-imagine connections of future technological and societal states of affair. These are produced by a variety of groups of actors of current society (e.g., developers, research policy makers, mass media). These futures are *socio*-technical as they are not limited to future technologies, but imagine and describe future socio-technical constellations, i.e. changes of social, political, legal, economic processes and structures coming along with new technologies. They are *socio-technical* as their generation and use is related to technological developments. To each respective technology we may attribute both structure-changing and structure-preserving effects. Socio-technical futures may refer to both unchanged, traditional technologies and to newly emerging ones. They always relate to technology, however the imagined technological developments may influence the imagined changes of societal situations to different degrees (e.g., technology as one factor out of many, such as with imaginations of climate change, e.g., technology as driving factor, such as with industrial robotics).

Imagined changes in the context of socio-technical futures may have different scales or range: socio-technical futures cover both widely shared "socio-technical imaginaries" (e.g. in the case of energy transformation or post-carbon society) and development- and innovation-guiding imaginations of the future which are limited e.g., to communities of engineers. However, they are also widely communicated visions, utopias, dystopias in the context of media discourses and/or scenarios from the political realm or governance contexts. Socio-technical futures are capable of describing, stating or proclaiming future changes of the entire society or of only some engineered sub-fields. They may refer to short-term or very far-reaching processes (e.g., a vision of a new product, a vision of a comprehensive reorganization of the world). Thus, socio-technical futures may be relevant for TA at its various and current places of activity, and in the context of its various processes (e.g., production processes, negotiation processes, e.g., laboratories, political arenas, media discourses). Particularly informative and instructive for TA are interactions and changes of these temporal and spatial dimensions of socio-technical futures.

2.2   In Which Ways and Where do Socio-Technical Futures Appear?

Socio-technical futures are found in many places of society. They are found in sciences, in research, development as well as in research policy and in the media. They appear in the context of communication and action processes of a variety of social actors-from major corporations as far as to new social movements. They play a role even for popular culture. Often they appear in plurality-even concerning the same topic-and proclaim different or even contradicting future states. Thus, socio-technical futures (or the solutions predicted by them) are not seldom the cause of controversies. These may be great public controversies, such as in the case of nuclear energy, but they may as well be limited to certain groups, such as the controversies among scientists in the early days of nanotechnology. For TA, socio-technical futures are relevant at all the places where they have an effect, due to their influence on current processes.

According to the great number of places where they have an effect, socio-technical futures may appear in very different ways, e.g., in the form of guiding visions of technologies, which are supposed to coordinate or provide orientation for research guidelines, or just for the practical work of a development department or project. They may also be part of products developed by help of scientific means, such as simulations, scenarios or roadmaps, which are explicitly created to provide orientation for decision-making and as a means of the further production of knowledge. Furthermore they may be formulated as long-term visions, for the purpose of raising the attention of certain communities, or of motivating actors to contribute to their realization. They are found in the context of utopias or dystopias structuring e.g. public controversies or motivating social movements. Socio-technical futures are always the products of heterogeneously distributed processes with actors from various fields of society (such as sciences, politics, business, law or art) contributing. All these kinds of socio-technical futures are relevant for TA if they are part of the technological developments and innovation processes the respective TA project is dealing with. However, socio-technical futures are not only formulated in the form of texts (such as research programs and the media) mentioning them; just the same they may be materially articulated in devices, images etc. or even organizations. Materials may explicitly stage socio-technical futures; just the same, certain socio-technical futures which were e.g., preferred by developers, may be implicitly inscribed into materials. TA should be capable of identifying all these kinds of socio-technical futures and should take them into consideration for analysis and assessment.

2.3   What is the Effect of Socio-Technical Futures in Present Times?

Socio-technical futures are efficacious in society, as they influence the actions, knowledge and decision-making of those actors who are dealing with them. This is due to the fact that they give expression to certain desires, fears, goals, interests, states of social groups or individuals, and to the way in which they see themselves. Due to prioritizing certain options, they have a feedback effect on the further development of the society and technology, as they present these options as being inevitable for solving current problems. Often socio-technical futures are subject to the rules of social "attention economy", they motivate actions, legitimate decisions and influence e.g., funding, regulation and the ways in which upcoming innovations are used and consumed. However, they may also be based on unquestioned and tacitly assumed ideas of the future (such as traffic always being characterized by the car). Also "incumbents" evading the attention of society have an action-motivating and decision-legitimating effect. Thus, socio-technical futures are always an expression of current states of affair and processes while in turn, contributing to shaping these states and processes at the same time. They reflect e.g., the predominance of certain constellations of actors and the matter-of-course nature of certain assumptions of the future. In the same way, they influence the course of processes of most different kinds (such as negotiations, production processes, regulations). TA must be capable of recognizing the ways in which they influence such processes and how they affect mentioned processes, as a precondition for critical assessment and, perhaps, to contribute to shaping them in a reflected way.

2.4   Why Should TA Deal Critically with Socio-Technical Futures?

Reflecting on converging and competing socio-technical futures enables TA to recognize both the plurality of future options and the limitations to certain options for the future. TA's critical dealing with socio-technical futures keeps us from considering those future changes as being predicted by one predominant socio-technical future, the sole promising option for the future. This way, an enlightened way of thinking in alternatives is supported. Each imagined socio-technical future necessarily prioritizes certain options at the expense of alternative possibilities. By making the ways and processes of this selection transparent,

TA keeps different options for future developments (e.g., "low-tech" or "no-tech" instead of "high-tech" solutions for current problems) open. Furthermore, the assessment of socio-technical futures by TA makes motivations, interests, ideas and expectations explicit, which have influenced their genesis and use. As a result of this "making explicit", they can be differentiated and negotiated. Their assessment is quite a crucial issue for TA, because they provide the basis on which innovation-relevant actors (try to) shape the future.

Furthermore, imagined socio-technical futures often give expression to desires and fears the concerned actors are not aware of. This is, among others, due to the fact that these imaginations are not tied to individual actors or limited groups of actors, but circulate among very different discourses and fields of society and may thus develop "a life of their own", which has hardly anything in common with the intentions of their producers. Therefore, socio-technical futures contain and develop their own efficacy in the context of communication and action processes, e.g. between citizens and mass media, between private organizations, enterprises, public institutions or even states and supra-national associations of states. The analysis of this efficacy helps TA to understand the backgrounds of social controversies and to make their implicit, basic assumptions analyzable and criticizable. By dealing with socio-technical futures, TA may gain insights concerning current social power constellations within the respective innovation or transformation context analyzed.

A comparison of different socio-technical futures (e.g., in different national and trans-national contexts) allows for identifying invariants and shared assumptions in the respective field of society. Looking at the ways in which socio-technical futures are used, points to in-/stabilities of social situations, to the in-/ exclusion of actors and their positions simply by virtue of selecting those technologies as being taken into consideration for the various socio-technical futures. Thereby, a differentiated analysis of the way socio-technical futures are being used, allows to uncover, question and assess the conditions for the selection and the construction of a problem and its solution. Narrow views in favor of technological options (such as nursing robots) while excluding or neglecting social solutions (such as a reorganization of nursing institutions) may be given as examples. By questioning these views TA takes a distance to narrations of "pure technological innovation". Such a critical view at technological innovations as social innovations is necessary for TA if it wants to analyze and assess the consequences of specific socio-technical futures in the context of innovation processes while, at the same time, taking care of, considering and directing its own position in the context of these processes (see Sect. 4).

3   How Could Socio-Technical Futures be Analyzed?

3.1   Which Analytical Dimensions Must be Distinguished?

Any analysis attempting to know in which ways current constellations and processes of society are included into socio-technical futures and, on the other hand, how these socio-technical futures change these constellations and processes, must try to understand the *mutual relation* of ideas about the future and current states of the society. For this purpose, each according to discursive and practice circumstances, contexts of making use of the various kinds of socio-technical futures have to be distinguished while their way of appearance has to be classified. Beyond this, it must be possible to record the interactions of forms and contexts. Thus, the following questions-among others-result: In which ways are innovation-guiding futures of engineers different from more general socio-technical futures such as visions, utopias or dystopias in the context of media discourses, and in which ways are these different from scenarios from the political realm and in governance contexts (see, places of effect and forms of socio-technical futures in Sect. 2.2)? And how could the interactions of the various kinds and contexts be analyzed?

For heuristic purposes, differentiating the contexts of the ways in which different kinds are used and recording the various interactions makes the distinction of two analytical dimensions reasonable. Only the inclusion of both dimensions allows to make statements on the significance and the effects of socio-technical futures. We take the analysis of how *"society finds expression in the futures"* as the first dimension; the second dimension is the analysis of the effect of *"futures in society"*. Here, society does not mean any totality such as "German society" or "globalized society" which can supposedly be reduced to one feature. Rather, "society" stands for those complex arrangements and dynamics of social actors that are being originated by actions, interests, power relations as well as negotiation and communication processes while, on the other hand, being characterized by established structures, norms, rules. Here, the society marks the respective context and process in the course of which socio-technical futures are produced, spread, used, controversially discussed.

The first analytical dimension focuses on finding out which current states of society find expression by a socio-technical future and in which way this leads to assumptions concerning future options that are desirable or undesirable or are considered realistic or unachievable. The second analytical dimension, on the other hand, focuses on grasping the effects (as well as the performativity) of a socio-technical future in its respective social context. Both dimensions must complement each other in the practice of research, otherwise the interactions of socio-technical futures and the contexts and processes of their production and use cannot be understood.

In the first dimension, the focus is on the respective socio-technical future as an object. It is about understanding what finds expression in a socio-technical future (e.g., a narrative, a vision, an image). Content and kind of this socio-technical future are analyzed in relation to its context. In the second dimension, the focus is on those social constellations and processes within which the respective socio-technical future has an effect. The effects of a specific socio-technical future can only be understood through the processes of its use, the discursive and actor constellations of its production, use, reception, modification etc. These are the starting points of socio-technical arrangements and their changes, the socio-technical future being just one element among others.

The efficacy (or also the performative character) of socio-technical futures can be examined by analyzing the content and nature of the respective socio-technical future in relation to the existing constellations and processes (e.g., production, use and inclusion into innovation and political processes). An analysis which focuses on the content and nature of a socio-technical future is an indispensable element for grasping e.g., normative preliminary assumptions or prioritizations of certain technological options. By reflecting on the context of its use, the in/exclusion of options become visible, a certain socio-technical future and the states of society expressed by it can be criticized. An analysis, which focuses on constellations and processes, grasps the dynamics of the consequences of socio-technical futures as an element of socio-technical arrangements. We reckon besides the technical ones also the economic, political, social and cultural structures among such arrangements.

3.2 What Must be Taken into Consideration for Deciding About the Analytical Method?

For the practice of research, both analytical dimensions must be connected to each other in a reflected manner. An analysis of "societies in the futures", which rather focuses on socio-technical futures as an object and on their topics and kinds (first dimension), allows for the information of various addressees and for the criticism of particular dimensions of socio-technical futures, such as their normative settings, their focus on a certain technological option and the connected exclusion of alternatives. An analysis of the actual effects (as well as the performative transformation power) of socio-technical futures as a part of socio-technical arrangements-that is "futures in society" (second dimension)-rather focusing on constellations and processes allows for an assessment and thus for criticism of the production, spread and use of certain socio-technical futures. Only such an empirical analysis allows for identifying socio-technical futures, which are momentous for innovation and transformation processes and are therefore relevant for TA.

The methods and procedures which might be applied are various and not at all new. Rather, there is a broad range of methodically structured analytical methods. For quite some time now, TA has been applying many different methods for its analysis of ideas about the future, such as vision assessment, scenario analysis and guideline assessment. According to what is necessary for a project, many of these theoretical concepts and methods are adopted and adjusted from other fields of practical work and sciences. Some of these fields might be STS, research on guiding visions, development of methods in sociology, philosophy, historical science, linguistic and cultural studies when it comes to the analysis of visionary discourses, collective expectations, models, symbols, metaphors, visual images, myths or also to the utopian contents of ideas of the future.

Basically, for the two dimensions of analyzing socio-technical futures, the same applies as for any reflected use of methods: the choice and combination of methods must be oriented towards the subject (e.g., the kind and nature of a socio-technical future), the goal of the respective study and its questions. According to the subject of "socio-technical futures", the qualitative methods of the humanities and the social sciences (e.g. from topical and discourse analyses via methods of field research as far as to constellation and process analyses) are in the fore. According to subject, other methods, such as from the cultural studies or economy, may be turned into a fruitful input for the analysis of socio-technical futures. Due to the complexity and variety of possible socio-technical futures (their kinds, places to have an effect, contexts etc.), the choice of methods requires reflected creativity.

4   How Could TA Situate Itself as an Actor of the Processes Under Analysis?

4.1   In Which Changing Contexts are Analyses Carried Out by TA?

TA's reflexive dealing with socio-technical futures requires a *contextualization of TA* within those constellations and processes of the current and changing society it is itself active in. Socio-technical futures have always been playing a role in the development and innovation processes TA deals with. They fulfil their tasks in the present, for example, as to organizing, adjusting, coordinating and communicating actor groups, that are dealing with a particular technological development. This means socio-technical futures have communication, coordination and motivation functions in the present. Accordingly, TA has been dealing with these functions already in the past, such as by its assessments of guiding visions. These days, however, a growing attention of and an increasing demand for socio-technical futures can be observed. This comes along with an expansion of far-reaching discourses on the future (concerning the chronological range and the spatial extent of statements on the future). This affects the self-location and self-reflection of TA, regardless whether it deals with far-reaching socio-technical futures or, as in the past, with closely and functionally limited socio-technical futures.

Thus, in many of the fields of technological development TA deals with, on the one hand an *expansion of far-reaching discourses on the future* can be observed. This does not only apply for NEST, but also for great transformations such as energy transformation, measures against climate change, big data etc. Technological future expectations are often not limited to expectations concerning functionality or the actual usefulness of individual technologies (such as, in the past, computer-aided-design-technology). Rather, socio-technical futures are communicated (in the form of visions, guiding principles, scenarios) which demand or require global, cross-generational changes (e.g., because they concern wide parts of society instead of only limited groups of users). Thus, these socio-technical futures describe much more than only technological innovations in the stricter sense, they promise radical change and transformations which might affect wide parts of the society. Thereby, these discourses on the future are far-reaching, not only due to the periods of time they address, but also when it comes to the consequences they envisage. On the other hand, this expansion comes along with politics, business, civil society and research paying *increasing attention* to socio-technical futures. In our current, technologized society, socio-technical futures are an essential element of the debate on the future of society.

Comprehensive innovation processes are made possible by the positions taken by very different social actors and their networks, in which TA itself is an actor. Its dealing with socio-technical futures plays a functional role in this context. This gets complimented by the fact that actors and addressees become more pluralized and differentiated by socio-technical futures and for the purpose of them. Constellations in the case of NEST (such as nanotechnology) or big data can serve here as examples, where socio-technical futures have the function of attracting attention, of negotiation, of coordination for the purpose of construing new fields of innovation to exploit great technological and economic potentials. Other constellations in which socio-technical futures become relevant and rather enable social movements are for example new ways of collective production (e.g., open source, open design) or the enabling and integration of real experiments including both experts and laypeople (such as the energy transformation). All these examples are about socio-technical constellations aiming at transformations, which fundamentally concern the society, both spatially and temporally. In this context, imagined futures play a crucial role as functional elements of the arrangements and processes of transformation; TA's dealing with socio-technical futures is always one element or practice of each overall constellation.

4.2   How Does Contextual Change Affect TA?

The expansion of discourses on the future and the growing attention of politics, business, research and the civil society correlates with a higher demand for visions and scenarios of future societal potentials of technology (e.g., when it comes to the socio-technical challenges of energy transformation). This becomes obvious by the increased activities of the strategic development of visions ("visioneering") and foresight methods, such as roadmaps and multi-dimensioned scenarios. In this changed context, TA's dealing with socio-technical futures becomes more important than before: on the one hand, TA is demanded to not only deal with limited socio-technical futures oriented at certain contexts (such as guiding visions for development), but also with so called far-reaching socio-technical futures. On the other hand, by its dealing with these futures, TA supports political guiding principles and programs. In this context, critical TA, is able to do critical agenda setting and to influence its context.

By way of its analyses and assessments of socio-technical futures, the research and advisory performance of TA-intended or unintended-contributes to modulating the discourses and practices of each respective context. Often this happens in competition with a variety of other actors and concerning unclear addressees of the respective TA expert opinion. These changed practice and intervention contexts of TA make obvious that it cannot play any non-situated, neutral role. Due to its dealing with the socio-technical futures of each respective context, TA obviously contributes to changing, thus modulating, these socio-technical futures and their consequences. If these influences are perceived as positive or negative, depends on each respective context and the self-reflective orientation of the respective TA practice.

The change of its contexts implies a more self-reflective concern by TA for its own role as an actor within the process while dealing with socio-technical futures. It also requests improved theoretical, methodical as well as empirical foundations (see Sect. 3). This requirements must be complied in order for TA to systematically understand and, accordingly, to critically orient its own statements on and assessments of socio-technical futures. For its clients and due to the growing awareness of socio-technical futures, TA as a service provider is demanded to identify more and more options for the future, to analyze and assess them, as well as to contribute to their production and to moderate processes of negotiating socio-technical futures. The clients expect TA to judge on long-term promises articulated in the form of visions. Whereas, spatially and chronologically far-reaching socio-technical futures have the effect that indeed not technology, but the society and its change becomes more than ever a topic and subject of TA (such as energy transformation, grand challenges). In this situation, TA can and must analyze socio-technical futures (incl. those, to whose shaping it contributes) as an expression and projection of current societal states. TA must be able to assess and criticize them concerning their consequences as well as their performative power when it comes to changes of those fields of society being affected by the relevant innovation processes.

4.3   How Does TA Situate Itself Between Support and Criticism of Socio-Technical Futures?

TA cannot avoid explicitly positioning itself-or being positioned by others-in the area of tension between critical assessment, on the one hand, and the support of certain socio-technical futures on the other hand. However, precisely this provides an opportunity, by way of critical assessment and a methodically-theoretically reflected dealing with socio-technical futures, by way of persistently emphasizing its current functions to point out exaggerated expectations and excluded alternatives as well as to question both power constellations and tacitly assumed normality. The task of such a TA of socio-technical futures is not only the assessment of technological functionalities and consequences, but informing about and criticizing societal processes as well as contributing to shaping them in a reflexive manner. This way, TA becomes an actor of the process who, by means of dealing with and insisting on socio-technical futures as an expression of current society and their effect on the development of the society, allows for a transparent assessment and comprehensible contribution to the shaping of futures. Most of all, such a present time-related dealing of TA with socio-technical futures protects from a thoughtless belief in certain socio-technical futures, as the only options for problem-solving and innovation. This position provides the basis for the reflected and critical moderation and also the co-shaping of those socio-technical futures, which are demanded under the conditions of each respective context.

From the implementations of its analyses about the current significance and function of socio-technical futures, result for TA, *as a research practice*, new knowledge desiderata. For classical TA, as the assessment of the consequences of limited technological developments, those desiderata were yet not necessary to this degree. Here, among others, comprehensive knowledge stocks of innovation research, STS, governance research, foresight studies and cultural studies have to be mentioned, whose inclusion must be achieved by increasing convergence and cooperation with the relevant sciences. The same applies for the implementation of the dimensions outlined under Sect. 3; it requires an improved integration of analytical knowledge and methods of linguistics, cultural studies and social sciences as well as philosophy to be able to empirically, functionally and critically analyze socio-technical futures by their current significance and effects, also in the context of extensive innovation and transformation processes. Furthermore, extensive practical skills are required in cases when the research practice of TA, as participatory and constructive TA (pTA; cTA), connects immediately to the practice of advice and shaping.

For *TA as an advisory practice* arise new demands to be able to derive advice-relevant knowledge from the knowledge of socio-technical futures; even more as there is a great number of potential addressees of TA expertise, from parliaments, scientists via civil society actors as far as to various publics. Accordingly, one must not only reflect on the question of which kind of knowledge is relevant for each advised actor, but also on the question of what might be the consequences of advisory practice in each respective constellation of actors. In this sense, TA opens up towards other kinds of expertise, also concerning its own production of knowledge, and welcomes the actors themselves to participate in the analysis process, not only as those seeking advice, but also as producers of knowledge. As pTA, it purposefully not only puts its knowledge and methodical expertise in the service of politics, but also enables civil society actors as well as the democratic public to reflect on socio-technical futures. With all its advice practices, however, TA must critically question which institutional, organizational, systemic, discursive, practical limitations the advisory practice in each respective constellation of innovation processes is subject to.

For *TA as a practice of the design* of socio-technical futures results that it must be more aware of its role as a modulating TA in the context of innovation and transformation processes. At the practical level, this implies that the analytics suggested by this paper must also be applied to its own practices of generating socio-technical futures (such as the development of scenarios, strategic visioneering in the context of foresight, integrated and process-accompanying development by cTA and pTA). This knowledge "of oneself" might serve TA for making a more effective and more strategic use of its critical expert opinions, thus situating itself within the multitude of future-related practices. The new attention economies in the changed contexts of innovation and transformation processes require the invention of new practices of advice and communication and thus of practices of critical shaping.

Recommended Readings

On Sections 1 and 2

Adam, B., & Groves, C. (2007). *Future matters: Action, knowledge, ethics.* Leiden: Brill.
Banse, G., Grunwald, A., Hronszky, I., & Nelson, G. (Eds.). (2011). *On prospective technology studies (KIT Scientific Reports 7599).* Karlsruhe: KIT Scientific Publishing.
Böhle, K. (2015). Desorientierung der TA oder Orientierungsgewinn? Einige Anmerkungen zum Vorschlag, die TA hermeneutisch zu erweitern. *Technikfolgenabschätzung-Theorie und Praxis, 24*(3), 91–97.
Brown, N., Rappert, B., & Webster, A. (Eds.). (2000). *Contested futures: A sociology of prospective techno-science.* Farnham: Ashgate.
Gransche, B. (2015). *Vorausschauendes Denken. Philosophie und Zukunftsforschung jenseits von Statistik und Kalkül.* Bielefeld: Transkript.
Grunwald, A. (2012). *Technikzukünfte als Medium von Zukunftsdebatten und Technikgestaltung.* Karlsruhe: KIT Scientific Publishing.
Grunwald, A. (2013). Techno-visionary sciences. Challenges to policy advice. *Science, Technology & Innovation Studies, 9*(2), 21–38.
Grunwald, A. (2015). Die hermeneutische Erweiterung der Technikfolgenabschätzung. *Technikfolgenabschätzung-Theorie und Praxis, 24*(2), 65–69.
Jasanoff, S., & Kim, S.-H. (Eds.). (2015). *Dreamscapes of modernity. Sociotechnical imaginaries and the fabrication of power.* Chicago: University of Chicago.
Nordmann, A. (2007). If and then: A critique of speculative nanoethics. *NanoEthics, 1*(1), 31–46.
Nordmann, A. (2010). A forensics of wishing: Technology assessment in the age of techno-science. *Poiesis & Praxis, 7*(1), 5–15.
Schaper-Rinkel, P. (2015). Antizipation von Zukunft zwischen Verwissenschaftlichung und Storytelling. In S. Azzouni, S. Böschen, & C. Reinhardt (Eds.), *Erzählung und Geltung. Wissenschaft zwischen Autorschaft und Autorität* (pp. 363–384). Weilerswist: Velbrück Wissenschaft.

Schulz, M. S. (2015). Special issue: Future moves in culture, society and technology. *Poiesis & Praxis, 63*(2), 129–139.

Selin, C. (2008). The sociology of the future. Tracing stories of technology and time. *Sociology Compass, 2*(6), 1878–1895.

Taylor, C. (2004). *Modern social imaginaries*. Durham: Duke University Press.

Torgersen, H. (2013). TA als hermeneutische Unternehmung. *Technikfolgenabschätzung-Theorie und Praxis, 22*(2), 75–80.

On Section 3

Alvial Palavicino, C. (2016). *Mindful anticipation. A practice approach to the study of emergent technologies*. Enschede: University of Twente.

Appadurai, A. (2013). *The future as cultural fact*. London: Verso.

Aykut, S. (2015). Energy futures from the social market economy to the Energiewende. The politicization of West German energy debates, 1950–1990. In J. Andersson & E. Rindzevičiūtė (Eds.), *Forging the future* (pp. 63–91). New York: Routledge.

Beckert, J. (2013). Imagined futures: Fictional expectations in the economy. *Theory and Society, 42*(3), 219–240.

Böhle, K., & Bopp, K. (2014). What a vision: The artificial companion. A piece of vision assessment including an expert survey. *Science, Technology & Innovation Studies, 10*(1), 155–186.

Dickel, S., & Schrape, J.-F. (2015). Dezentralisierung, Demokratisierung, Emanzipation. *Zur Architektur des digitalen Technikutopismus. Leviathan, 43*(3), 442–463.

Dieckhoff, C., Appelrath, H.-J., Fischedick, M., Grunwald, A., Höffler, F., Mayer, C., et al. (2014). *Zur Interpretation von Energieszenarien. Schriftenreihe Energiesysteme der Zukunft*. München: acatech-Deutsche Akademie der Technikwissenschaften e. V.

Dierkes, M., Hoffman, U., & Marz, L. (1992). *Leitbild und Technik. Zur Entstehung und Steuerung technischer Innovationen*. Berlin: Edition Sigma.

Geideck, S., & Liebert, W.-A. (Eds.). (2003). *Sinnformeln. Linguistische und soziologische Analysen von Leitbildern, Metaphern und anderen kollektiven Orientierungsmustern*. Berlin: De Gruyter.

Grin, J., & Grunwald, A. (Eds.). (2000). *Vision assessment: Shaping technology in 21st century society. Towards a repertoire for technology assessment*. New York: Springer.

Konrad, K., Markard, J., Ruef, A., & Truffer, B. (2012). Strategic responses to fuel cell hype and disappointment. *Technological Forecasting and Social Change, 79*(6), 1084–1098.

Leitner, K.-H., Warnke, P., & Rhomberg, W. (2016). New forms of innovation. Critical issues for future pathways. *Foresight, 18*(3), 224–237.

Levitas, R. (2013). *Utopia as method: The imaginary reconstitution of society*. New York: Springer.

Lösch, A. (2013). Vision Assessment zu Human-Enhancement-Technologien. Konzeptionelle Überlegungen zu einer Analytik von Visionen im Kontext gesellschaftlicher Kommunikationsprozesse. *Technikfolgenabschätzung-Theorie und Praxis, 22*(1), 9–16.

Lösch, A. (2014). *Die diskursive Konstruktion einer Technowissenschaft. Wissenssoziologische Analytik am Beispiel der Nanotechnologie*. Baden-Baden: Nomos.

Lösch, A., & Schneider, C. (2016). Transforming power/knowledge apparatuses: The smart grid in the German energy transition. *Innovation: The European Journal of Social Science Research, 29*(3), 262–284.

Scheer, D. (2013). *Computersimulationen in politischen Entscheidungsprozessen: Zur Politikrelevanz von Simulationswissen am Beispiel der CO2-Speicherung*. Wiesbaden: Springer VS.

Scheer, D., & Renn, O. (2014). Public perception of geoengineering and its consequences for public debate. *Climatic Change, 125*(3–4), 305–318.

Schulz-Schaeffer, I. (2013). Scenarios as patterns of orientation in technology development and technology assessment. Outline of a research program. *Science, Technology & Innovation Studies, 9*(1), 23–44.

Schulz-Schaeffer, I., & Meister, M. (2015). How situational scenarios guide technology development-Some insights from research on ubiquitous computing. In D. M. Bowman, A. Dijkstra, C. Fautz, J. Guivant, K. Konrad, H. van Lente, & S. Woll (Eds.), *Practices of innovation and responsibility. Insights from methods, governance and action* (pp. 165–179). Heidelberg: AKA/IOS.

te Kulve, H., Konrad, K., Alvial Palavicino, C., & Walhout, B. (2013). Context matters: Promises and concerns regarding nanotechnologies for water and food applications. *NanoEthics, 7*(1), 17–27.

Warnke, P., & Schirrmeister, E. (2016). Small seeds for grand challenges—Exploring disregarded seeds of change in a foresight process for RTI policy. *Futures, 77*, 1–10.

Wentland, A. (2016). Imagining and enacting the future of the German energy transition: Electric vehicles as grid infrastructure. *Innovation: The European Journal of Social Science Research, 29*(3), 285–302.

On Section 4

Ahlqvist, T., & Rhisiart, M. (2015). Emerging pathways for critical futures research: Changing contexts and impacts of social theory. *Futures, 71*, 91–104.

Coenen, C., & Simakova, E. (2013). STS policy interactions, technology assessment and the governance of technovisionary sciences. *Science, Technology & Innovation Studies, 9*(2), 3–20.

Dickel, S. (2013). Die Regulierung der Zukunft. "Emerging Technologies" und das Problem der Exklusion des Spekulativen. In A. Bora, A. Henkel, & C. Reinhardt (Eds.), *Wissensregulierung und Regulierungswissen* (pp. 201–218). Weilerswist: Velbrück.

Dieckhoff, C. (2015). *Modellierte Zukunft-Energieszenarien in der wissenschaftlichen Politikberatung*. Bielefeld: Transkript.

Grunwald, A. (2008). *Technik und Politikberatung. Philosophische Perspektiven*. Frankfurt a. M.: Suhrkamp.

Haraway, D. J. (1997). *Modest_Witness@Second_Millennium. FemaleMan_Meets_OncoMouse: Feminism and Technoscience*. New York: Routledge.

Konrad, K., Stegmaier, P., Rip, A., & Kuhlmann, S. (2014). Constructive technology assessment-Antizipation Modulieren als Teil der Governance von Innovation. In M. Löw (Ed.), *Vielfalt und Zusammenhalt. Verhandlungen des 36. Kongresses der Deutschen Gesellschaft für Soziologie in Bochum und Dortmund 2012*. Frankfurt a. M.: Campus (CD-Rom).

Konrad, K., van Lente, H., Groves, C., & Selin, C. (2016). Performing and governing the future in science and technology. In C. A. Miller, U. Felt, R. Fouché, & L. Smith-Doerr (Eds.), *The handbook of science and technology studies* (4th ed., pp. 465–493). Cambridge: MIT Press.

Latour, B. (2004). Why has critique run out of steam? From matters of fact to matters of concern. *Critical Inquiry, 30*(2), 225–248.

McCray, W. P. (2012). *The visioneers: How a group of elite scientists pursued space colonies, nanotechnologies, and a limitless future*. Princeton: Princeton University Press.

Nordmann, A. (2013). Visioneering assessment: On the construction of tunnel visions for technovisionary research and policy. *Science, Technology & Innovation Studies, 9*(2), 89–94.

Nordmann, A. (2014). Responsible innovation, the art and craft of anticipation. *Journal of Responsible Innovation, 1*(1), 87–98.

Poli, R. (2014). Anticipation: What about turning the human and social sciences upside down? *Futures, 64*, 15–18.

Rip, A. (2012). Futures of technology assessment. In M. Decker, A. Grunwald, & M. Knapp (Eds.), *Der Systemblick auf Innovation. Technikfolgenabschätzung in der Technikgestaltung* (pp. 29–42). Berlin: Sigma.

Schaper-Rinkel, P. (2006). Governance von Zukunftsversprechen: Zur politischen Ökonomie der Nanotechnologie. *PROKLA, 145*, 473–496.

Schaper-Rinkel, P. (2013). The role of future-oriented technology analysis in the governance of emerging technologies: The example of nanotechnology. *Technological Forecasting and Social Change, 80*(3), 444–452.

Schneider, C., & Lösch, A. (2015). What about your futures, Technology Assessment? An Essay on how to take the visions of TA seriously, motivated by the PACITA conference. *Technikfolgenabschätzung-Theorie und Praxis, 24*(2), 70–74.

Schot, J., & Rip, A. (1996). The past and future of constructive technology assessment. *Technological Forecasting and Social Change, 54*, 251–268.

Schot, J., & Steinmüller, W. E. (2016). *Framing innovation policy for transformative change: Innovation policy 3.0*. Brighton: University of Sussex. http://www.johanschot. com/publications/framing-innovation-policy-for-transformative-change-innovation-policy-3-0/.

Suchman, L., & Bishop, L. (2000). Problematizing 'Innovation' as a critical project. *Technology Analysis & Strategic Management, 12*(3), 327–333.

Weber, K. M., Amanatidou, E., Erdmann, L., & Nieminen, M. (2016). Research and innovation futures. Exploring new ways of doing and organizing knowledge creation. *Foresight, 18*(3), 193–203.

Andreas Lösch (Ph.D. and habilitation) is sociologist, senior researcher and head of the research area "knowledge society and knowledge policy" and of the research group on "vision assessment" at the Institute for Technology Assessment and Systems Analysis at the Karlsruhe Institute of Technology.

Knud Böhle is sociologist and information scientist specialized in Technology Assessment of ICT and foresight, researcher at the Institute for Technology Assessment and Systems Analysis at the Karlsruhe Institute of Technology since 1986.

Christopher Coenen is political scientist and senior researcher at the Institute for Technology Assessment and Systems Analysis at the Karlsruhe Institute of Technology. His work focusses on socio-cultural-political aspects of new and emerging technosciences in areas of overlap between biology and technology.

Paulina Dobroc is linguist and semiotician and PhD student at Institute for Technology Assessment and Systems Analysis and at Institute for German Studies at Karlsruhe Institute of Technology and scholarship holder of Hans Böckler Foundation.

Arianna Ferrari (Ph. D.) is philosopher and head of the Department Strategy and Content at Futurium gGmbh, a new space in Berlin, which works as astage, museum, laboratory and forum of the future (www.futurium.de).

Reinhard Heil is philosopher and researcher at the Institute for Technology Assessment and Systems Analysis at the Karlsruhe Institute of Technology.

Dirk Hommrich (Ph.D.) is philosopher and Scientific Officer for International Developments at the Head Office of the German Council for Scientific Information Infrastructures (RfII), Göttingen.

Martin Sand (Ph.D.) is philosopher of technology and a Marie Skłodowska-Curie Fellow at the Department of Values, Technology and Innovation at TU Delft, where he works on moral luck in science and innovation.

Christoph Schneider (PhD.) is a sociologist with focus on equitable and democratic digital futures, 3D printing and Responsible Research and Innovation at the Institute for Technology Asessment and Systems Analysis at the Karlsruhe Institute of Technology.

Stefan C. Aykut (Ph.D.) is assistant professor (Juniorprofessor) of sociology at the department of socioeconomics of the University of Hamburg, and associated member of LISIS, Université Paris-Est MLV.

Sascha Dickel (Ph. D.) is sociologist and assistant professor (Juniorprofessor) for sociology of media at Johannes Gutenberg University Mainz. He is head of the collaborative project "Prototyping Futures", funded by the Federal Ministry of Education and Research.

Daniela Fuchs is human ecologist and junior researcher at the Institute of Technology Assessment of the Austrian Academy of Sciences in Vienna.

Bruno Gransche (Ph.D.) has been a philosopher at the Institute of Advanced Studies (FoKoS) at the University of Siegen since 2017. He works in the fields of philosophy of technology, ethics, and future-oriented thinking. He is a research fellow at the Fraunhofer ISI in Karlsruhe, where he worked as a philosopher and Foresight expert until 2016.

Armin Grunwald (Ph.D. and habilitation) is Director of the Institute for Technology Assessment and Systems Analysis and Full Professor of Philosophy and Ethics of Technology at the Karlsruhe Institute of Technology. Simultaneously, he heads the Office of Technology Assessment at the German Bundestag.

Alexandra Hausstein (Ph.D.) is sociologist, researcher at and managing director of the Institute of Technology Futures at Karlsruhe Institute of Technology (KIT).

Karen Kastenhofer (Ph.D.) is a biologist and scholar in Science and Technology Studies (STS) at the Institute of Technology Assessment of the Austrian Academy of Sciences in Vienna.

Kornelia Konrad (Ph.D.) is Assistant Professor of Anticipation and Assessment of Emerging Technologies in the Department of Science, Technology and Policy Studies (STePS) at the University of Twente, NL.

Alfred Nordmann (Ph.D.) is Professor of Philosophy and History of Science and Technoscience at Darmstadt Technical University and an adjunct professor in the Department of Philosophy at the University of South Carolina. He explores the formation of conceptions of scientific objectivity as well as epistemological, metaphysical, aesthetic aspects of technoscientific research.

Petra Schaper-Rinkel (Ph.D.) is political scientist and Senior Scientist at the Center Innovation Systems & Policy at Austrian Institute of Technology. She works in the field of Innovation Dynamics and Modelling with the research focus on Foresight, Innovation Policy, Science and Technology Studies.

Dirk Scheer (Ph.D.) is senior research scientist at the Institute for Technology Assessment and Systems Analysis at the Karlsruhe Institute of Technology. His research focus is on social-science based energy research and technology acceptance, knowledge transfer and management at the science-policy interface, and studies on participation and risk.

Ingo Schulz-Schaeffer (Ph.D. and habilitation) is Full Professor and head of the sociology of technology and innovation group at the Institute of Sociology, Technical University of Berlin. He is principal investigator of the DFG cluster of excellence "science of intelligence", and member of the program committee of the DFG priority program "digitalization of working worlds".

Helge Torgersen (Ph.D.) is a biologist by education, has been working on GMO safety regulation, biotechnology policy and the relation between science, technology and society at the Institute of Technology Assessment of the Austrian Academy of Sciences in Vienna.

Alexander Wentland (Ph.D.) is a postdoctoral researcher at the Munich Center for Technology in Society (MCTS) in the Innovation, Society & Public Policy group. His work focuses on sociotechnical imaginaries of sustainability and regional diversity in innovation governance.